STUDY PROCESSES HANDBOOK

FIRST EDITION | MONOGRAPH 35

Published by:
The Australasian Institute of Mining and Metallurgy,
Ground Floor, 204 Lygon Street, Carlton Victoria 3053, Australia

All chapters published in this volume were peer reviewed prior to publication.

ISBN 978 1 922395 37 5

Desktop published by Kristy Burt
Cover design by Brooke Mills
The Australasian Institute of Mining and Metallurgy

FOREWORD

The purpose of this handbook is to provide our members with a guide to the minimum requirements for resource project studies, providing consistency to the discipline of study processes and reports.

In recent years there has been warranted increased scrutiny on the competence of the technical work on which business cases supporting forward-looking statements have been presented to stakeholders. It is highlighted that this handbook is not a guide on how to conduct studies but instead presenting the framework with minimum standards. The objective is to enhance the professional capability of our members and to underpin investor and community confidence, corporate governance, and the general reputation in the mineral resources sector. The handbook guidelines encourage the consistent application of good practice minimum standards against which stakeholders and practitioners alike can assess the quality/integrity of the study on which informed decisions are made.

Successful delivery of resource projects is intrinsically linked to the progression through the study phases (Scoping, Pre-Feasibility, Feasibility) and into final design, construction and operation. The definition of success means different things to different stakeholders through the study/project life cycle. The study reports should present to stakeholders a business case with relevant information including the value proposition as well as identifying the material risks and opportunities presented by a project to enable informed decisions at each phase of assessment.

The framework this handbook has adopted is to provide the minimum standards of each phase of study (Scoping, Pre-Feasibility, Feasibility) by technical discipline. Level of accuracy is the key driver to inform the extent to which each discipline needs to achieve for these phases. From experience not all disciplines can achieve the level of accuracy at the same time within a study phase; however, transparent reporting with a forward plan for the next phase of study is acceptable. This is not acceptable where there is a potential fatal flaw not addressed or where material residual risk impact on the outcomes of the project is not addressed. Such instances may be thought of as the 'weak links' in the process and the study accuracy may well be constrained by such flaws or unresolved risks.

It is noted that the discipline of the study process has been open to interpretation by practitioners and the definition of what constitutes good practice varies significantly across the industry and jurisdictions.

The concept for this handbook was the precursor to the AusIMM Study Processes for Resource Projects online course, first presented in 2022, evolving from discussions in the AusIMM Consultants Society during 2020/21. The Study Processes subcommittee and authors comprise a group of experienced subject matter experts for each of the disciplines to contribute to the contents of this handbook.

The handbook is a guide for study practitioners and contributors to address the relevant aspects at the level appropriate for the phase of a study. The focus of the handbook is:

- doing the *right work* at the right time
- doing the *right things first* before you try to do them properly
- asking if you are building the *right project* in the first place.

The handbook does not cover *how* to do the work, as this is presented in other publications and texts, including AusIMM monographs and conference proceedings, for example, Monograph 27 *Cost Estimation Handbook* and Monograph 30 *Mineral Resource and Ore Reserve Estimation – The AusIMM Guide to Good Practice*. For this reason, the contents of the Study Processes Handbook serve as a valuable resource to complement Reporting Standards (the JORC Code and the VALMIN Code) and Corporate Disclosure Guidelines (ASX Listing Rules). The handbook specifically applies AusIMM terminology but it is equally applicable to other jurisdictions such as CIM, SAIMM, and others.

This handbook provides a useful and practical overview and description of the requirements for the phases of the study process that support the assessment, evaluation, and valuation of mineral projects.

Professor Mark Noppé FAusIMM(CP)
Director, WH Bryan Mining Geology Research Centre
Sustainable Minerals Institute
The University of Queensland

CONTRIBUTORS

AusIMM wishes to thank the following contributors for lending their expertise and their time to this project.

CHAPTER 1 – INTRODUCTION

Peter Fairfield FAusIMM(CP), Principal Consultant, Miner Insight Pty Ltd

Paul Harper FAusIMM, CEO and Principal Consultant, Enthalpy Pty Ltd

Lisa Park FAusIMM(CP), Principal, Process Engineering Options

Karl van Olden FAusIMM, Executive Lead – Operations, AMC Consultants

Irena Ivanova FAusIMM, General Manager Evaluation Studies, Neometals Ltd

CHAPTER 2 – ENVIRONMENT, SOCIAL AND PERMITTING

Alan Irving FAusIMM(CP), Principal Environmental Consultant, Alan Irving and Associates

Bruce Harvey FAusIMM(CP), Geologist, Social Performance Professional

CHAPTER 3 – GEOLOGY AND RESOURCES

Sonia Konopa FAusIMM, Manager Resources, Principal Geologist, ERM

Matthew Nimmo, Principal Geoscience and Data Scientist, AMC Consultants

CHAPTER 4 – MINING AND RESERVES

Glen Williamson FAusIMM(CP), Operations Consulting Manager Perth, AMC Consultants

CHAPTER 5 – METALLURGY

Lisa Park FAusIMM(CP), Principal, Process Engineering Options

CHAPTER 6 – PROCESS PLANT AND INFRASTRUCTURE

Lisa Park FAusIMM(CP), Principal, Process Engineering Options

Irena Ivanova FAusIMM, General Manager Evaluation Studies, Neometals Ltd

CHAPTER 7 – TAILINGS STORAGE AND WATER MANAGEMENT

Tim Rowles FAusIMM(CP), Regional Manager, Knight Piésold Consulting
Arjuna Dissanayake, Senior Engineer, Knight Piésold Consulting
Andrew Brown, Senior Engineer, Knight Piésold Consulting
Dr Shehan Udukumburage, Senior Geotechnical Engineer, Knight Piésold Consulting
Peter Jenner, Head of Tailings (Zinc Group), Glencore International AG

CHAPTER 8 – PROJECT DEVELOPMENT PLAN

Adrian Pratt FAusIMM(CP), Principal, Adrian Pratt Mining Consultants

CHAPTER 9 – COST ESTIMATION

Damian Peachey, Principal Consultant, AMC Consultants
Karl van Olden FAusIMM, Executive Lead – Operations, AMC Consultants

CHAPTER 10 – ECONOMIC EVALUATION

Ashton Lombardo, Director, Corporate Finance, BDO
Sherif Andrawes, Partner, Global Head of Natural Resources, BDO
Adam Myers, Partner, BDO

CHAPTER 11 – RISK

Paul Peter, Risk Management Specialist, Auxilium One

SHERIF ANDRAWES

Partner, Global Head of Natural Resources, BDO

Sherif Andrawes serves as BDO's Head of Global Natural Resources and Energy, encompassing mining, oil and gas, and renewables, leveraging over 35 years of experience in business valuations, mergers and acquisitions, and independent expert reports. With extensive valuation experience spanning various industries with a focus on the mining sector, Sherif has authored over 600 independent expert reports, solidifying his reputation as a leader in Australia's financial sector.

Recognised as a Fellow of Chartered Accountants Australia and New Zealand (CAANZ) and the Institute of Chartered Accountants in England and Wales, as well as a CA BV (Business Valuation) Specialist as recognised by CAANZ, Sherif has contributed significantly to the field through presentations and publications throughout his career.

ANDREW BROWN

Senior Engineer, Knight Piésold Consulting

Andrew Brown joined Knight Piésold in 2012 after graduating from Monash University in Melbourne with a Bachelor of Environmental Engineer (Hons) in 2011. Andrew has over 12 years of experience in tailings management, climatology, hydrology, flood modelling, and dam break assessments.

Andrew has broad ranging involvement in tailings and water dam design and management in Australia, South-East Asia, East Africa and West Africa. Andrew currently heads the climate, hydrology and hydraulic design team providing technical direction, review and management within the Perth office of Knight Piésold Pty Ltd.

ARJUNA DISSANAYAKE

Senior Engineer, Knight Piésold Consulting

Arjuna Dissanayake graduated from the University of Peradeniya, Sri Lanka with a Bachelor of Science in Engineering in 2001. Arjuna has more than 22 years of experience in civil, structural, geotechnical designs in mining projects. Arjuna has undertaken projects from conceptual design through to detailed design and construction of tailings management systems, water dams, waste dumps and other mine site infrastructure.

Arjuna is Engineer of Record for a 190 m high extreme hazard earth/rockfill dam and a 90 m high hazard tailings dam in Southeast Asia and is Deputy Engineer of Record for several tailings dams in Australia. Arjuna is a Chartered Professional Engineer and a Registered Professional Engineer in Queensland in the field of Civil Engineering.

PETER FAIRFIELD FAusIMM(CP)

Principal Consultant, Miner Insight Pty Ltd

Peter is a process-driven, strategically minded mining engineer with 35 years of experience in operations management and consulting providing technical and operational support. After 20 years in site-based operational roles Peter has over 17 years' experience working on and leading multi-discipline study teams on projects across commodities and regions. He has extensive experience in multi-discipline project evaluations, independent technical reviews, and due diligence projects, and has managed and prepared independent reports for internal and public reporting and disclosure.

PAUL HARPER FAusIMM

CEO and Principal Consultant, Enthalpy Pty Ltd

Paul is a mining engineer with over 40 years' experience and is currently CEO at Enthalpy Pty Ltd. During his consultancy career Paul has led numerous Independent Peer Reviews and feasibility studies. He has also fulfilled several senior roles including Global Head of Mining Anglo American, CEO AMC Consultants, Mining Practice Lead Advisian, Operations Manager Clough Engineering, Mine Manager at Northparkes and Renison Tin Mines, and Chief Mining Engineer at Mt Lyell Copper Mine.

BRUCE HARVEY FAusIMM(CP)

Geologist, Social Performance Professional

Bruce has 45 years of worldwide field experience in the resources sector, first as an exploration geologist and for the past 25 years as a social performance professional. He was Global Practice Leader Social Performance at a major resources company for eight years. He is an Adjunct Professor at the Sustainable Minerals Institute at the University of Queensland.

As a member of the AusIMM Social and Environment Committee, he facilitated the development of the Institute's Social Responsibility Framework and the Environment and Social Performance competency development tool. He led the development of the AusIMM Professional Certificate in ESG and Social Responsibility and the Operationalising ESG short course.

ALAN IRVING FAusIMM(CP)

Principal Environmental Consultant, Alan Irving and Associates

Alan Irving has had a long career in the mining industry since starting out as a forester working in mine rehabilitation. He has worked on-site in coal operations (Queensland and Indonesia) and copper/gold operations (Western Australia and Indonesia). He later specialised in environmental impact assessment and feasibility studies and is now an independent environmental consultant. He has worked on projects in Indonesia, Malaysia, Peru, Paraguay, Brazil, Mozambique, Mongolia, UAE, New Caledonia and Australia.

IRENA IVANOVA FAusIMM

General Manager Evaluation Studies, Neometals Ltd

Irena holds a Master's degree in Chemical Engineering and has worked in the minerals and metals industry for over 25 years. Her commodity experience covers copper, nickel, cobalt, uranium, vanadium, magnetite / hematite, gold and battery recycling materials. Irena has extensive expertise in process design, technology implementation, study, project and engineering management and team development, gained through her roles in international engineering houses, such as Hatch, Fluor, Tetra Tech and, most recently, Ausenco, as a General Manager Minerals and Metals for APAC / Africa.

PETER JENNER

Head of Tailings (Zinc Group), Glencore International AG

Peter Jenner is Head of Tailings for Glencore Zinc and is based in Baar, Switzerland, with over 30 years of industry experience. Peter is experienced in corporate governance for tailings storage facilities including auditing and preparation of tailings standards and other governance documents. He also has practical experience in tailings design and management and has significant experience for both operators and consultants.

As well as being highly skilled in tailings governance, management and construction Peter has a great deal of experience in the geotechnical and geological assessment and appraisal of sites for the mining, energy, water/wastewater and industrial sectors. He has carried out numerous feasibility and design level studies in the mining, water/wastewater, infrastructure, energy and oil and gas sectors and has taken several of these through to the construction and operational phase. Peter's experience is truly global, having worked and managed many projects around the world in over 50 countries on six continents.

Peter has a BSc (Hons) in Applied Geology from Staffordshire (1994). He is a Fellow of the Geological Society of London and is a Certified Natural Scientist with SACNASP (Sth Africa).

SONIA KONOPA FAusIMM

Manager Resources, Principal Geologist, ERM

Sonia is a resource geologist with over 35 years of international experience in the mining industry. She has previously worked in various operational and leadership roles across Australia, Asia Pacific, and Europe, and has extensive international expertise in consulting services, technical advice and guidance across a range of commodities and geological settings. Most recently, Ms Konopa has held resource management roles at the Martabe Gold Mine and Toka Tindung Gold Mine in Indonesia. Her broad practical experience extends to Mineral Resource estimation, exploration, project management, and corporate transaction advisory consulting services.

ASHTON LOMBARDO

Director, Corporate Finance, BDO

Ashton Lombardo is a finance and accounting professional with over 12 years of corporate finance experience preparing valuations and independent expert reports, investigating accountants' reports, and conducting due diligence across various industries, specialising in natural resources. Ashton has recently been appointed to the VALMIN Committee, where he provides valuation and corporate and regulatory insight.

Ashton holds a Bachelor of Commerce and a Bachelor of Economics degree from the University of Western Australia. He is also a member of the Institute of Chartered Accountants and a Business Valuation Specialist. He holds a Graduate Diploma in Applied Corporate Governance from the Governance Institute of Australia.

ADAM MYERS

Partner, BDO

Adam Myers is a seasoned Corporate Finance Partner within the BDO Advisory team, specialising in Natural Resources. With over 26 years of experience, Adam's expertise encompasses valuations, expert reports, and financial modelling. Adam's extensive repertoire includes advising on IPOs, conducting due diligence, and providing strategic counsel on international transactions.

Adam holds a Bachelor of Commerce in Accounting and Finance and a Graduate Diploma in Applied Finance and Investment. He is a Fellow of Chartered Accountants Australia and New Zealand and also serves as a Non-Executive Member of the JORC Committee, further attesting to his standing as a distinguished professional in the field.

MATTHEW NIMMO

Principal Geoscience and Data Scientist, AMC Consultants

Matthew is a geologist with 30 years of experience in the mining industry as a project and resource geologist. For the first 20 years Matthew focused on mineral resource estimation and statistical modelling of geological data. For the last ten years Matthew has shifted focus to working as a data scientist in the mining industry where he uses his domain expertise combined with data analytics, machine learning, and programming skills to extract insights from and model geology, geotechnical, geometallurgy, hydrology and equipment maintenance data.

Matthew's expertise includes data wrangling, data analytics, data visualisation, machine learning, statistical hypothesis testing, spatio-temporal modelling, simulations (Monte-Carlo and geostatistical), and numeric optimisation. Matthew is skilled at using the R language and environment for data analysis and automation of data analysis.

LISA PARK FAusIMM(CP)

Principal, Process Engineering Options

Lisa is a technical specialist (metallurgy) and project manager with more than 25 years in the resources industry. Lisa's experience spans projects in multiple continents, from exploration to delivery. She is committed to identifying, assessing and mitigating risks to lead teams safely and achieve critical objectives.

DAMIAN PEACHEY

Principal Consultant, AMC Consultants

Damian has more than 40 years of experience in the mining industry that includes broad operational and technical roles in management, maintenance, engineering, procurement and logistics. He has held a range of senior management roles in iron ore operations in Australia.

His focus in recent years with AMC has been on performance diagnostic studies, mining equipment asset management, selection and procurement, and cost estimations for global open-pit mining operations.

PAUL PETER

Risk Management Specialist, Auxilium One

Paul is a risk management consultant with over 25 years international experience across different disciples and industries. He was awarded the CIO Advisor Magazine Top Enterprise Risk Management Consultant of the year in 2020.

Over the last five years, Paul has assisted companies to develop and implement risk frameworks, policies and procedures, reviewed organisation's risk management programs and specialises in the facilitation of risk assessment workshops (including Option Analysis, HAZID, HAZOP, SFAIRP, FMEA and Bow-tie Analysis).

Paul previously held positions of Regional Audit Manager and Head of Risk and Governance for global energy and resources companies. He has worked in major capital projects, commodity trading and treasury risk, and business strategy across the USA, Chile, Africa, Singapore, Netherlands and Australia.

ADRIAN PRATT FAusIMM(CP)

Principal, Adrian Pratt Mining Consultants

Adrian Pratt has a diverse background in mining projects that spans their evaluation, design, implementation, and operation. In 2014 he established his own consultancy, which focuses on mineral project development in Australia and internationally. Prior to this he held senior technical support and operational roles with Ivanhoe Australia, Newcrest, BHP and Normandy Mining.

Adrian holds a Bachelor of Engineering (Mining) from the University of NSW and a Masters of Minerals and Energy Economics from Macquarie University, and Certificates of Competency as a Mine Manager. He is a Fellow of the Australasian Institute for Mining and Metallurgy with Certified Professional (CP) accreditation, a Member of the Institute of Engineers Australia with registration as a Certified Practicing Engineer and is a Registered Professional Engineer of Queensland (RPEQ).

TIM ROWLES FAusIMM(CP)

Regional Manager, Knight Piésold Consulting

Tim has over 25 years of experience in engineering and environmental studies for mining projects. His efforts have included studies (concept through to detailed design), environmental impact assessments, construction, and implementation for tailings management systems, mine water management systems, water dams, waste rock dumps, mine site infrastructure and mine closure.

Tim has project experience across Australia as well as site and design experience on projects located in the Pacific, Asia, Europe, South America and Africa, including for base metal, precious metal, rare earth, energy metals and bulk commodity projects. Tim is on the Independent Technical Review Boards for several high hazard tailings dams and is engineer of record (and deputy engineer of record) for a suite of tailings dams ranging from large upstream paddock facilities in arid/semi-arid regions to high (>100 m) downstream earthfill and rockfill tailings dams in tropical monsoonal climates.

Tim completed his undergraduate degree at the Royal School of Mines at Imperial College, London (1996), prior to undertaking a master's degree at the University of Manchester (1998). He is a Fellow and Chartered Professional with AusIMM, a member of the Australian Institute of Geoscientists and a Registered Professional Engineer of Queensland in the field of environmental engineering.

DR SHEHAN UDUKUMBURAGE

Senior Geotechnical Engineer, Knight Piésold Consulting

Shehan Udukumburage is a senior geotechnical engineer with ten years of experience in mining and civil engineering projects including three years of post-graduate studies in geotechnical engineering. Shehan's experience in geotechnical engineering includes site investigations, foundation design, tailing dam design, seepage and slope stability, ground improvement and numerical modelling of geotechnical problems.

Shehan has published a number of technical papers on the laboratory based investigation methods for expansive clays and won the best paper award for one of his conference papers published in 8th International Conference on Geotechnique held in Malaysia (2018). He has been involved in a diverse range of projects as a design engineer from concept to detailed design stages. Furthermore, he has demonstrated ability in the application of geotechnical design tools and programs to model complex subsurface ground conditions to assess the geotechnical slope stability and design retaining structures.

KARL VAN OLDEN FAusIMM

Executive Lead – Operations, AMC Consultants

Karl has more than 30 years' mining industry experience in diverse mining environments in operational, technical, consulting, and corporate roles. He has expertise and skills across the mining business process and is a trusted advisor to executives in the mining industry. He has been a project sponsor, a project manager, a technical lead and a contributor to well over 500 diverse, world-wide mining industry assignments. These span feasibility studies, strategic analysis, business and life of mine planning, Ore Reserve estimates, due diligence reviews, expert reports, minerals project valuations and operational optimisation.

GLEN WILLIAMSON FAusIMM(CP)

Operations Consulting Manager Perth, AMC Consultants

Glen has over 40 years of experience in the mining industry in a range of commodities, and in roles ranging from mine planning, operations management, technical management, feasibility studies, due diligence reviews, and Ore Reserve estimation and reporting. He has worked for mining companies, mining contractors and in consulting. Glen has worked on open pit iron ore, manganese, coal, nickel laterite, bauxite, phosphate, potash, precious metals, and base metals projects throughout Australia, the Pacific region, south-east Asia, South America, Africa, Russia, and Kazakhstan and has been Competent Person/Qualified Person under the JORC Code, NI 43-101, and S-K 1300 and has reviewed Ore Reserves reported under the SAMREC Code.

STEERING COMMITTEE

Peter Fairfield FAusIMM(CP)
Paul Harper FAusIMM(CP)
Irena Ivanova FAusIMM
Lisa Park FAusIMM(CP)
Karl van Olden FAusIMM

REVIEWERS

AusIMM wishes to thank the following reviewers for their contribution to this publication.

Diane Dowdell FAusIMM
Ian Gordon
Simon Hanrahan FAusIMM(CP)
Stephen Kittel
Michelle Lawson FAusIMM
Peter McCarthy HonFAusIMM(CP)
Mark Noppé FAusIMM(CP)
George Widelski

CONTENTS

1

INTRODUCTION

ABOUT THE HANDBOOK

This handbook has been prepared to offer improvements in the consistency, quality and content of studies for resource projects.

It seeks to provide greater transparency on the status of studies through the application of a recommended framework. It is hoped that this approach will improve the reputation of the resources industry and strengthen relations with community, stakeholders and investors in a world that is seeking full transparency of key risks and opportunities.

The purpose of this handbook is to provide a high-level framework for study processes. This will provide a consistent approach against set ethical standards, and accountability through a set of guidelines to which stakeholders can refer. Ultimately these guidelines aim to enhance the deliverable outcomes of capital projects to improve industry performance and increase investor and community confidence.

The organising committee offered this framework to subject matter experts to present the information in each chapter. The framework is based on material that is currently available in the public arena and supports other Australasian Institute of Mining and Metallurgy (AusIMM) and industry publications.

The handbook is a supporting document for the Study Processes for Resource Projects online course facilitated by AusIMM.

This handbook is not a reporting standard. While there are specific requirements for external reporting of Mineral Resources and Ore (Mineral) Reserves, there are no specific disclosure guidelines or requirements for reporting study results, which include technical and financial assessments. By way of example, Canada's National Instrument (NI) 43-101 Reporting Framework is for reporting the results of technical studies and is not a framework for what is to be addressed in each study phase.

In the committee's experience, the reporting framework has become the focus of the study rather than the work required within the study, regardless of the study phase. What all stakeholders require is full transparency of the business case (value proposition and risk profile) being put forward for approval.

CONTEXT

The objective of developing this handbook is to provide frameworks and detail prepared by experienced industry practitioners that can be used by study managers and teams to manage the process by which resource projects are advanced through the study phases.

From the committee's experience, the lack of prescribed definitions of study contents has resulted in a wide interpretation of technical study assessment. This has led to debates about the standard and quality of the findings and reporting of technical studies. This is particularly evident in the assessment and diligence investigations undertaken for and by potential project investors.

The framework is provided to enable a continual process of risk management, commencing at the Scoping Study phase, through to completion of the Feasibility Study phase (Figure 1.1).

It is important to first define the purpose of a study and then ensure this is appropriately addressed. The purposes may include:

- identifying the potential value in a project in order to request funding to advance the study
- determining the risk profile pertaining to a project and what needs to be addressed in the next phase of work
- understanding the external factors that a project may impact upon with the environment, communities and regulators
- applying alternative technologies to greenfield or brownfield projects
- expanding a brownfield project
- supporting a funding strategy to develop and execute a project
- assisting in a major company merger or acquisition.

Studies are typically undertaken for internal company purposes, which is the focus of this handbook. The study work is then, where required, converted to a format suitable for the public domain.

The purpose of this handbook is to increase understanding of what may be considered a minimum technical requirement to be addressed, from internal and external company, investor, financier and regulator perspectives. The handbook does not attempt to provide guidance on how to perform technical work or a 'reporting structure' for various regulatory and corporate structures.

This handbook seeks to complement the Australasian Code for Reporting of Exploration Results, Mineral Resources and Ore Reserves (JORC Code, 2012), the Australasian Code for Public Reporting of Technical Assessment and Valuations of Mineral Assets (VALMIN Code, 2015), and stock exchange reporting and disclosure requirements across the various regulatory authorities.

TARGET AUDIENCE

The target audience for this handbook is practitioners who manage and participate in studies on resource projects and those decision-makers and leadership teams that approve these studies.

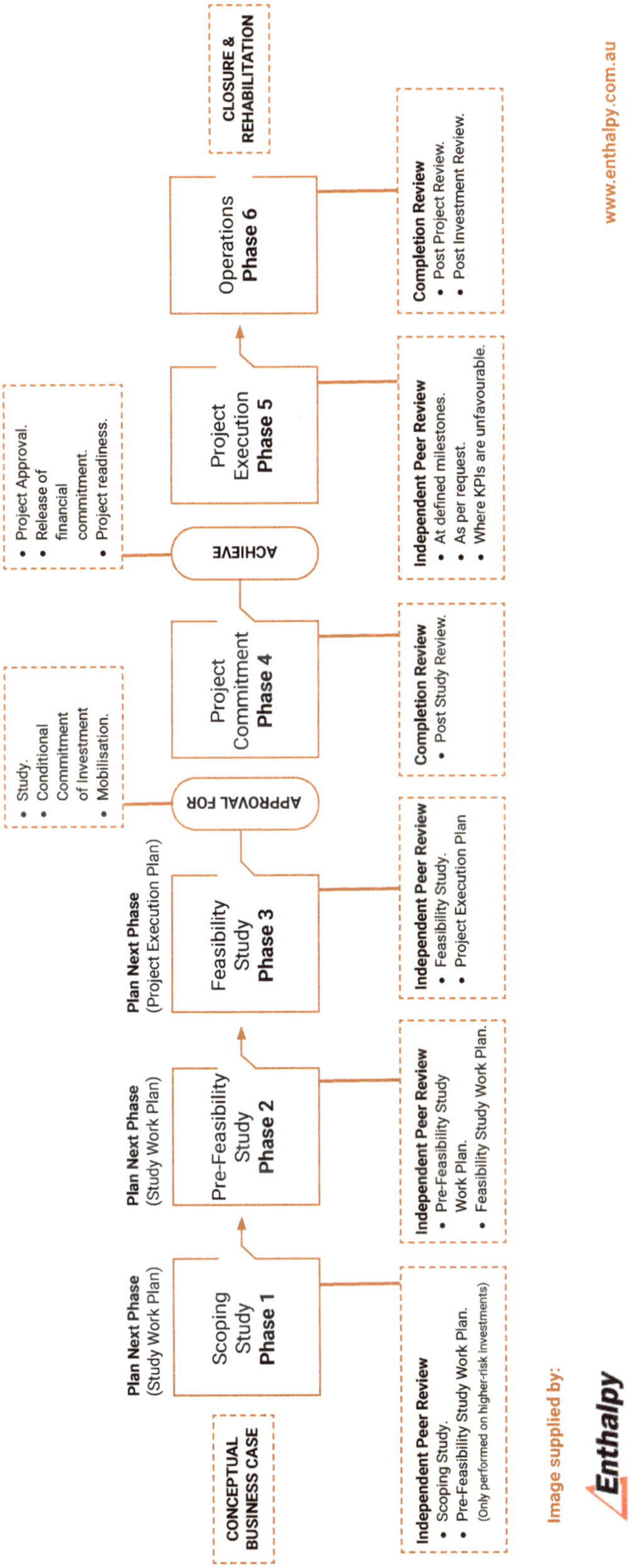

FIG 1.1 – The six phases of resource project studies.

STRUCTURE OF THE HANDBOOK

The handbook is intentionally structured by 'discipline' as shown in the high-level framework (Figure 1.2) rather than defining requirements for a Scoping, Pre-Feasibility and Feasibility Study. This enables the relevant specialist to consider the work required within the discipline and provide the appropriate level of understanding to support the necessary level of confidence for technical risk, cost estimation and economic assessments.

The intent is to have a consistent approach with minimum requirements (level of detail and accuracy) for each discipline at each phase of the study.

The objectives of each phase of study are as follows:

- *Scoping Study* – Is there a potential project and if so, is it worth investing funds to look at options for further evaluation? The study needs to present the value versus risk and whether the project is a good strategic fit for stakeholders.
- *Pre-Feasibility Study* – Options to develop and execute the project are evaluated with the aim of selecting one option to advance to the next phase of work. This part of the study cycle is generally the most challenging because this is when opportunities and risks of a project are better understood. In most cases, time and funding constraints to deliver a Pre-Feasibility Study are the enemy.
- *Feasibility Study* – The single option determined from the Pre-Feasibility Study is fully evaluated. The study needs to present, in full detail, how to optimise the project and mitigate risk. It should also contain the expected financial outcomes. In addition, the execution and operational strategies need to be fully understood by the relevant decision-makers to ensure that the project is viable.

SUMMARY

In conclusion, this handbook contains guidelines to help relevant stakeholders understand the study phases and processes. The aims are to assist stakeholders to progress through the study process and execute capital projects more effectively, and to encourage decision-makers to exercise robust due diligence and accountability, as required by good governance.

The intent is to lift the standard of studies being delivered. This will in turn enhance the reputation of the resources industry by increasing investor confidence and improving community–regulator relations, while minimising unnecessary impact on the environment.

REFERENCES

JORC, 2012. Australasian Code for Reporting of Exploration Results, Mineral Resources and Ore Reserves (The JORC Code) [online]. Available from: <http://www.jorc.org> (The Joint Ore Reserves Committee of the Australasian Institute of Mining and Metallurgy, Australian Institute of Geoscientists and Minerals Council of Australia).

VALMIN Committee, 2015. Australasian Code for Public Reporting of Technical Assessments and Valuations of Mineral Assets (The VALMIN Code) [online]. Available from: <https://www.valmin.org/docs/VALMIN_Code_2015_final.pdf>. (The VALMIN Committee, a joint committee of the Australasian Institute of Mining and Metallurgy and Australian Institute of Geoscientists).

	SCOPING STUDY	PRE-FEASIBILITY STUDY (PFS)	FEASIBILITY STUDY (FS)
INTRODUCTION	Scene setting/basis for study. Basic maps and property claims.	Scene setting/basis for study. Preliminary maps and property claims.	Scene setting/basis for study. Detailed maps and property claims.
ENVIRONMENT, SOCIAL AND PERMITTING	Conceptual. Permitting pathway and requirements defined. Reference to international, national requirements.	Data collection, monitoring program to international, national requirements.	Detailed and ongoing, ESIA, permitting to international, national requirements.
DEPOSIT DEFINITION	Limited drilling, sampling and assaying.	Preliminary drilling, sampling and assaying.	Defined/extensive drilling, sampling and assaying.
RESOURCES AND RESERVES	Predominantly Inferred Resources. Reserves not reported.	Predominantly Indicated Resources. Typically Probable Reserves.	Indicated and Measured Resources. Probable and Proved Reserves.
MINING (Open pit / Underground / Other)	Assumptions. Simplified designs.	Methodology identified. General equipment selection. Preliminary mine design.	Methodology selected. Mine planning optimised. Equipment selected. Geotechnical investigations.
METALLURGY (Sampling / Test work)	Limited definition of geometallurgy. Process benchmarking to inform to recovery and throughput.	Preliminary geometallurgy definition. Preliminary ore characterisation, bench scale test work, trade-off studies.	Detailed geometallurgy definition. Sample selection across major and minor domains. Detailed variability test work. Piloting for novel flowsheets. Confirmed recovery and throughput.
ENGINEERING	Preliminary processing plant flowsheet. Estimates, assumptions. <5% discipline engineering.	Processing plant flowsheet options assessed. Preliminary engineering methods. Preliminary studies, power and water identified. 5–25% engineering.	Detailed processing plant flowsheet, pre-Engineering Design. Power and water confirmed. 20–50% engineering.
PROJECT DEVELOPMENT PLAN	General, estimates based on assumptions.	Preliminary.	Detailed.
CAPITAL COST ESTIMATE	Accuracy -25%+50%, Contingency ±30%	Accuracy -15%+25%, Contingency 15–30%	Accuracy -10%+15%, Contingency 10–15%
OPERATING COST ESTIMATE	Accuracy -25%+50%, Contingency ±25%	Accuracy -15%+25%, Contingency ±15%	Accuracy -10%+15%, Contingency ±10%
ECONOMIC EVALUATION	Assessment, budget quotes, simple analysis.	Preliminary analysis.	Detailed assessment and analysis.
RISK	General overview.	Fatal flaw, risk workshop.	Fatal flaw, risk workshop, simulations.

FIG 1.2 – The study processes matrix.

ENVIRONMENT, SOCIAL AND PERMITTING

2

INTRODUCTION

The focus on environment, social and permitting aspects of mining and mineral processing continues to intensify and evolve over time. Access to Mineral Resources and the ability to develop them successfully requires that a clear and rigorous understanding of these aspects be developed through all study phases of proposed projects.

APPROVALS, PERMITS, LICENCES AND CONSENTS

A project may need to comply with and obtain the following (Table 2.1):

- tenure and land access
- overarching regulatory context
- individual approvals, permits, licences and consents.

TABLE 2.1

Environment, social and permitting study phase guide.

ENVIRONMENT, SOCIAL AND PERMITTING					
Activity	**Deliverable**	**Description**	**SS**	**PFS**	**FS**
Land access	Ownership and tenure reports	Tenure and ownership – identification, access, acquisition	✓	✓	✓
		Mineral and other – identification, access, acquisition	✓	✓	✓
	Agreements			✓	✓
	Procedures			✓	✓
Approvals, permits and licences	Reports	Stakeholder engagement plan	✓	✓	✓
		Permitting plan	✓	✓	✓
		Environmental and social impact assessment		✓	✓
		Risk assessment	✓	✓	✓
		Closure plan	✓	✓	✓
Planning	Schedule		✓	✓	✓
	Budget		✓	✓	✓

SS = Scoping Study; PFS = Pre-Feasibility Study; FS = Feasibility Study.

Tenure and land access

In the Scoping Study phase, the study team is expected to:

- Identify existing tenure and existing ownership details of all areas where access may be required to develop the project. Include likely locations for all future field investigations proposed in the Pre-Feasibility Study, not just exploration.
- Identify easements, rights-of-way, protected areas, sites of cultural heritage significance and land subject to Indigenous land claims or existing customary usage land rights.
- Obtain mineral and other relevant tenure for all proposed activities requiring on-site investigations. This includes, as applicable in the local jurisdiction, access for exploration, bulk sampling or mining, pilot plants, and infrastructure (including access roads, pipelines, power lines, port facilities).
- Commence obtaining associated land access rights or make agreements with landholders and Indigenous people (as required).
- Commence discussions with relevant land-connected Indigenous groups towards a free, prior and informed consent (FPIC) agreement (as required).
- Put in place a tenement and agreement tracking and reporting system with assigned accountabilities.

As the study moves into Pre-Feasibility, the study team should:

- Prepare a detailed and integrated schedule of all land purchases, mineral tenements, leases, easements and rights-of-way required for each option under consideration.
- Identify any joint venture or heads of agreement matters that relate to tenure or land access matters and understand the implications thereof.
- Establish land access protocols for field investigations.
- Commence land purchase negotiations where required.
- Commence making agreements with land-connected Indigenous groups that will demonstrate FPIC is in place (where relevant).
- Commence all tenure application process(s) required by regulators.
- Update the tenement and agreement tracking and reporting system, including identification of critical path activities.

In Feasibility, the study team should:

- Obtain all mineral-related, infrastructure and other tenures for construction and operations.
- Finalise land purchases and access agreements as required.
- Ensure there is a high degree of certainty of obtaining any outstanding tenure in a timely manner and the pathway is documented and risks known.
- Update the tenement and agreement tracking and reporting system.

Regulatory context

In the Scoping Study phase, the study team is expected to:

- Understand the key legislation, regulations and by-laws governing development projects and the departments/agencies responsible for their administration.
- Identify factors relating to the regulatory context that add complexity and/or uncertainty to the development of the project (eg capacity of regulators, health and safety, product quality, hazardous wastes, employment and contract law, environmental and socio-political aspects, overlapping regulations).

As the study moves into Pre-Feasibility, the study team should ensure:

- Pending and potential changes to legislation and regulations are monitored and assessed for impact on the project development schedule, capital and operating cost estimates.
- International intergovernmental agreements and conventions, which may affect regulations in the jurisdiction in which the project is located, are monitored and assessed as above.

In Feasibility, the focus changes to tracking:

- Pending and potential changes to legislation and regulations are monitored and assessed for impact on the project development schedule, capital and operating cost estimates.
- International intergovernmental agreements and conventions, which may affect regulations in the jurisdiction in which the project is located, are monitored and assessed as above.

Approvals, permits, licences and consents

The Scoping Study phase focuses on identification. The study team should:

- Identify the approvals, permits, licences and consents required for each option under consideration. This includes, but is not limited to, planning and environment, tenure and land access, ancillary infrastructure, labour regulations and foreign investment rules.
- Identify any fatal flaws in options under consideration (eg potential sustained opposition from regulators and/or other stakeholders, very high risk to schedule, risk of successful third-party legal challenge).
- Obtain site access approval for drilling and other early investigatory works.
- Scope studies required in the Pre-Feasibility Study to support applications for approvals, permits, licences and consents (including from land-connected Indigenous groups and other natural resource users, as required) and prepare budget estimate.

The Pre-Feasibility phase includes:

- Preparing a detailed integrated schedule of all approvals, permits, licences and consents required for each option under consideration. Identification of critical path items. Benchmark timelines against analogous projects in the same jurisdiction.
- Commencing formal environmental and social impact assessment (ESIA) process(es) required by regulators. Where more than one project option is being evaluated, ensure all options are able to be covered under the scope of the ESIA regulatory process.
- Scheduling and resourcing ESIA and other studies to meet defined regulatory milestones.
- Briefing regulators about the project so that they understand and agree to the approvals pathway. Regulators include local government, state/provincial and national governments, as required.

By the end of Feasibility:

- ESIA studies required by regulators, and others such as financial institutions, should be complete.
- All primary approvals, permits, licences and consent agreements have been obtained for construction and operations. Should there be any outstanding approvals, permits, licences or consents, ensure there is a high degree of certainty of obtaining them in a timely manner and the pathway is documented and risks known.

- Secondary approvals required as conditions of primary approvals (eg environmental management and monitoring plans, biodiversity offset plans) have been obtained for construction and are well advanced for operations.
- A compliance tracking and reporting system is in place with assigned accountabilities.

ENVIRONMENT

Baseline information

In the Scoping Study phase, the study team is expected to:

- Identify project-related environmental aspects where collection of new baseline data is a priority (eg vegetation and fauna surveys, hydrology, hydrogeology, meteorology, noise, air quality) and where seasonal and multi-year data are necessary.
- Scope baseline data collection and schedule to meet project study milestones and ESIA requirements.
- Establish weather stations, monitoring bores, gauging stations and other long-term monitoring systems, where required, to meet the overall project schedule.

During Pre-Feasibility, the study team should:

- Acquire a full range of baseline environmental information required to:
  - help decision-making regarding site selection, mining method, location of infrastructure and mine-related disturbance
  - support the ESIA.
- Coordinate biophysical monitoring programs and environmental surveys with other study-related field work (eg resource drilling, geotechnical investigations, on-ground and aerial surveying, community surveys).

During Feasibility, the study team should:

- Largely complete baseline data collection for ESIA purposes.
- Continue data collection where:
  - there is uncertainty about the regulator's view of adequacy of data supporting the ESIA
  - seasonal and multi-year data is needed to improve model accuracy (eg groundwater model, flood model)
  - data from reference sites will be needed to assess construction and operational impacts.

Risk assessment

In the Scoping Study phase, the study team is expected to:

- Identify potential environmental threats, including threats to gaining regulatory approvals.
- Evaluate threat significance for each project option and identify possible mitigation measures.
- Identify potential fatal flaws and potential opportunities.
- Use outcomes of risk assessment to help scope baseline information collection, scope impact assessment studies and formulate regulatory approvals strategy.

In the Pre-Feasibility Study phase, the study team is expected to:

- Undertake further detailed risk assessment of project options.
- Identify and detail mitigation measures to reduce the likelihood and/or consequences of significant risks.

- Use the outcomes of the risk assessment to help scope baseline information collection and impact assessment studies, and select the preferred option.

In the Feasibility Study phase, the study team is expected to:

- Finalise a detailed risk assessment of the preferred project option.
- Refine mitigation measures to reduce likelihood and/or consequences of significant risks.
- Refine measures to take advantage of identified opportunities.

Design criteria

In the Scoping Study phase:

- Identify environmental aspects that are likely to require the setting of design criteria for the project (eg noise, vibration, emissions to air and water, solid and liquid wastes, flooding).
- Identify and collate existing design and receiving environment limits adopted by regulators in the local jurisdiction (including national and international standards) or set by the proponent's corporate standards.
- Identify opportunities to apply waste hierarchy, circular economy and eco-efficiency principles in design.

During the Pre-Feasibility Study phase, the study team should:

- Obtain relevant technical specifications from suppliers of plant and equipment (eg sound power levels, quality of water treatment plant output).
- Set initial environmental design criteria for environment aspects identified in the Scoping Study.
- Include relevant design criteria in modelling required as part of ESIA studies (eg modelling of noise, air dispersion, floods).
- Assess opportunities to apply waste hierarchy, circular economy and eco-efficiency principles in design.
- Identify any potential constraints to meeting proposed design criteria during operations and consequential cost and technical implications.

The Feasibility phase should:

- Finalise environmental design criteria and include specifications for suppliers and contractors.
- Document any design criteria imposed as conditions of regulatory approvals, permits and licences and changes necessitated by results from impact assessments.
- Clearly communicate any changes to the engineering design and procurement teams.

Environmental and social impact assessment

The environmental and social impact assessment (ESIA) approval is typically the principal approval required for a project, from which specific construction and operational permits and licences are then derived.

During the Scoping Study phase:

- Identify environmental features of importance that may be affected by the project (eg watercourses, protected species, high-value habitat, cropping land, clean airshed, residences). Carry out preliminary characterisation of physical and geochemical properties of soil, waste rock, tailings and other process residues.

- Make a preliminary assessment of likely environmental impacts, identify knowledge gaps, and prepare a detailed scope and estimate the cost of impact assessment work to be undertaken in the Pre-Feasibility Study. Use environmental risk assessment to help define scope.
- Evaluate options under consideration in terms of likely environmental acceptability. Identify potential fatal flaws and potential opportunities.

In the Pre-Feasibility Study phase:

- Carry out a detailed ESIA in accordance with:
  - study scope
  - regulatory ESIA requirements
  - requirements for permit and licence applications.
- Evaluate options under consideration in terms of environmental acceptability. Identify fatal flaws and potential opportunities.
- Prepare a detailed scope and estimate the cost of impact assessment work to be undertaken in the Feasibility Study. Use the updated environmental risk assessment to help define scope.

The outcomes of the Feasibility Study phase include:

- Completed and approved ESIA studies required by regulators.
- Environmental management and monitoring plans based on the updated environmental risk assessment and ESIA outcomes.

Environmental management and monitoring

In the Scoping Study phase, the study team is expected to:

- Identify key environmental management measures required for the options under consideration.
- Identify key environmental monitoring requirements needed to support the ESIA.
- Prepare high-level estimates of capital expenditure (CAPEX) and operating expense (OPEX) impact.

In the Pre-Feasibility Study phase, the study team is expected to:

- Prepare preliminary environmental management and monitoring plans for construction and operations (consistent with requirements of the ESIA and applications for regulatory approvals).
- Ensure environmental management measures are aligned with environment design criteria and relevant CAPEX and OPEX included in the project financial model.
- Set up an environmental data management system.

The outcomes of the Feasibility Study phase include:

- Environmental management and monitoring plans finalised for construction and operations (consistent with the ESIA, conditions of regulatory approvals and proponent's environmental management system).
- Environmental management measures are aligned with environment design criteria and relevant CAPEX and OPEX included in project financial model.
- A compliance tracking and reporting system is in place with assigned accountabilities.
- A fully maintained environmental data management system.

Energy and greenhouse gas emissions

In the Scoping Study phase, the study team is expected to:

- Make a preliminary assessment of energy demand and assess energy supply options, including low greenhouse gas (GHG) emission energy sources (eg wind, solar, hydroelectricity, hydrogen, nuclear).
- Make a preliminary estimate of annual demand for each proposed energy source (eg electricity, diesel, gas, fuel oil, coal, biomass).
- Make a preliminary estimate of requirements for on-site storage (eg diesel, gas, batteries, hydrogen) to ensure reliability of supply.
- Make a preliminary estimate of annual scope 1, scope 2 and material scope 3 GHG emissions. Identify potential for GHG offsets and carbon capture and storage. Identify the potential impact of project emissions on any existing or proposed company GHG emissions targets.
- Identify energy-related and GHG-related regulatory requirements (existing or pending) that may influence GHG emissions, energy supply and energy reliability. Prepare high-level estimate of CAPEX and OPEX impact.

In the Pre-Feasibility Study phase, the study team is expected to:

- Quantify energy demand and assess energy supply options, including low GHG emission energy sources (eg wind, solar, hydrogen). Estimate annual demand for each proposed energy source (eg electricity, diesel, gas, fuel oil, coal, biomass). Quantify requirement for on-site storage (eg diesel, gas, batteries, hydrogen) to ensure reliability of supply.
- Quantify annual scope 1, scope 2 and material scope 3 GHG emissions for each option being considered. Assess potential for GHG offsets and carbon capture and storage. Identify potential impact of project emissions on any existing or proposed company GHG emissions targets.
- Identify energy-related and GHG-related regulatory requirements (existing or pending) that may influence GHG emissions, energy supply and energy reliability. Quantify CAPEX and OPEX impact.

The Feasibility Study phase should:

- Finalise energy demand and detail the adopted energy supply approach.
- Quantify annual demand for each proposed energy source (eg electricity, diesel, gas, fuel oil, coal, biomass). Quantify requirement for on-site storage (eg diesel, gas, batteries, hydrogen) to ensure reliability of supply.
- Quantify annual scope 1, scope 2 and material scope 3 GHG emissions. If proposed, describe GHG offsets or carbon capture and storage details. Quantify impact of project emissions on any existing or proposed company GHG emissions targets.
- Identify energy-related and GHG-related regulatory requirements (existing or pending) that may influence GHG emissions, energy supply and energy reliability. Quantify CAPEX and OPEX impact.

Climate change

In the Scoping Study phase, the study team should:

- Gather existing information on any projected changes to regional climate (eg rainfall intensity and amount, drought, cyclone frequency, temperature, evaporation) to help

inform development of initial hydrological and hydrogeological models, and a water supply assessment.
- Consider climate change adaption threats and opportunities in risk assessments.

During the Pre-Feasibility Study phase, climate change assessments should:

- Include projected climate change trends and weather extremes in hydrological and hydrogeological models and water supply assessments of options under consideration.
- Update climate change adaption threats and opportunities in risk assessments and describe mitigation measures.

In Feasibility, assessments should:

- Include projected climate change trends and weather extremes in numerical hydrological and hydrogeological models and the water supply assessment.
- Update climate change adaption threats and opportunities in risk assessments. Detail and cost mitigation measures.

Closure planning

General closure planning considerations are set out below. Specific social closure planning considerations are presented in the 'Social closure planning' section at the end of this chapter.

The Scoping Study phase should:

- Identify closure-related regulatory requirements, including requirements of the ESIA.
- Identify significant closure-related risks for each option under consideration.
- Identify possible post-closure land uses and repurposing options and how they may differ for each project option under consideration.
- Prepare high-level closure cost estimates (including cash flow) for options under consideration for inclusion in project financial models.

During Pre-Feasibility, the study should:

- Identify closure-related regulatory requirements, including requirements of the ESIA.
- Describe outcomes of any stakeholder consultation about post-closure land use.
- Identify significant closure-related risks and liabilities for each option under consideration.
- Describe preliminary closure objectives and targets and indicative completion criteria for proposed post-closure land uses and how they may differ for options under consideration.
- Prepare high-level closure cost estimates (including cash flow) for options under consideration for inclusion in project financial models.

The Feasibility phase should:

- Identify closure-related regulatory requirements, including requirements of the ESIA.
- Describe outcomes of closure-related stakeholder consultation.
- Detail closure objectives and targets and indicative completion criteria for preferred post-closure land use.
- Identify significant closure-related risks and describe mitigation measures to be undertaken during operations.
- Finalise the closure cost estimate (including cash flow) for the preferred option for inclusion in the project financial model.

SOCIAL

Social and economic baseline

The Scoping Study phase should:

- Identify and gather existing social and economic baseline information.
- Complete initial field appraisal by subject matter experts of formal and informal land use rights, including the identification of any legally recognised Indigenous land connections and customary heritage rights.
- Identify project-related social and economic aspects where collection of new baseline data is a priority. Understand seasonal and multi-year usage of existing natural resources (eg land, water, forage).
- Scope baseline data collection and schedule to meet project study milestones and ESIA requirements. Commence long-term data gathering where required to meet overall project schedule.

During the Pre-Feasibility Study phase, the study team should acquire the full range of baseline social and economic information required to:

- Support the ESIA.
- Help decision-making regarding site selection, mining method, location of infrastructure and project-related disturbance.
- Assess future accommodation and common-user infrastructure (roads, transmission lines, airfields, etc) requirements to support various project options.
- Assess capability of local suppliers of goods and services and assess local labour supply and workforce capability.

The Feasibility phase should:

- Largely complete the baseline data collection for ESIA purposes.
- Maintain up-to-date baseline information during the Feasibility Study to enable assessment of future construction and operational impacts.
- Identify requirements for government/civic/landowner partnership agreements or consent agreements needed to support project development and operations.
- Establish a social and economic data management system.

Social risk assessment

During the Scoping Study, the study team should:

- Identify potential social threats, including threats to gaining regulatory approvals and 'social licence to operate'.
- Evaluate threat and opportunity significance for possible project options and identify possible threat mitigation and opportunity enhancement measures.
- Identify any potential fatal flaws and 'no regrets' opportunities.
- Use outcomes of risk assessment to help scope baseline information collection and impact assessment studies, and to formulate stakeholder consultation and engagement strategy.

In the Pre-Feasibility Study phase, the study team should:

- Enhance the detailed social risk assessment of project options based on early engagement with stakeholders.

- Identify and detail mitigation measures to reduce likelihood and/or consequence of significant social threats. Evaluate opportunities for enhancing social support.
- Use outcomes of social risk assessment to help scope baseline information collection and social impact assessment studies, and evaluate options under consideration.
- Establish a social induction program and codes of conduct for on-ground Feasibility work that can roll though to construction and operations.

During Feasibility, the study team should:

- Continue to update social impact assessment and risk assessment to help define final design.
- Finalise the detailed risk (threat and opportunity) assessment of the preferred project option.
- Refine mitigation measures to reduce the likelihood and/or consequences of significant threats.
- Refine measures to take advantage of identified opportunities to enhance social support for preferred option.

Community consultation and engagement

During Scoping, the study team should:

- Undertake stakeholder mapping and analysis.
- Prepare the stakeholder consultation and engagement plan.
- Prepare initial public information materials.
- Consider potential large social impacts, and discuss with key informants potential opportunities (eg partnerships) for mitigation.
- Design community concerns, complaints and grievance procedure and document initial concerns from stakeholder and key informants.
- Open dialogue with any legally recognised land-connected Indigenous groups regarding potential partnering and agreement expectations, including about FPIC.

During Pre-Feasibility:

- Undertake consultation with affected community groups and conduct targeted stakeholder engagement to support the ESIA. Establish and make publicly accessible a community concerns, complaints and grievance procedure.
- Identity affected community group concerns with the options under consideration and engage on issues that have significant project development threats or opportunities.
- Engage with affected community groups and targeted stakeholders on significant risks to help optimise project design to lower threat of social opposition. Document outcomes.
- Identify potential opportunities (eg partnerships) that will enhance 'social licence to operate' and document scope, schedule, budget and accountable parties (eg potential partners).
- Commence agreement making with land-connected Indigenous groups as required.

In Feasibility, the study team is expected to:

- Establish a project stakeholder engagement coordination committee.
- Specifically engage with affected communities and targeted stakeholders to support the ESIA and potential project development pathway.

- Identity community residual concerns with the preferred option and specifically engage on issues that have significant project development risks.
- Use engagement on significant risks to help optimise project design for public acceptability to lower risk. Document outcomes.
- Identify opportunities (eg partnerships) to mitigate social risk and document scope, schedule, budget and accountable parties.
- Update community and stakeholder consultation and engagement plan in preparation for construction. Update public information materials.
- Finalise Indigenous consent agreements that demonstrate FPIC as required.

Cultural heritage management

During the Scoping Study, the study team should:

- Identify and gather existing public domain cultural heritage information.
- Consult with legal and custodial landowners, their representative bodies and other relevant parties regarding land access.
- Undertake preliminary cultural heritage reconnaissance, inclusive of tangible, intangible and geological/palaeontological features.
- Scope detailed cultural heritage surveys and schedule them to meet project study milestones and ESIA requirements.
- Prepare cultural heritage protocols for all field investigations and land disturbance activities during studies.

In Pre-Feasibility, the study team should:

- Undertake cultural heritage surveys to:
  - support the ESIA
  - help decision-making regarding site selection, location of project and associated shared use infrastructure and project-related disturbance.
- Prepare a cultural heritage management plan, including workforce induction approach, where necessary.
- Maintain cultural heritage protocols including induction and 'chance find' procedure for all field investigations and land disturbance during studies.

During Feasibility, the study team should:

- Finalise any remaining cultural heritage surveys.
- Update the cultural heritage management plan including site access protocols for custodians, and naming and celebration events where desired.
- Prepare cultural heritage protocols for land disturbance activities during construction, formalised into consent agreements meeting FPIC requirements where required.

Social impact assessment

During the Scoping Study phase, the study team should:

- Confirm preliminary characterisation of the social and economic environment obtained from the baseline assessment and use this to identify possible social impacts, including possible impacts on human rights, public health, safety and security.
- Make a preliminary assessment of likely social and economic impacts, identify knowledge gaps and prepare detailed scope and cost of impact assessment work to be undertaken in

the Pre-Feasibility Study. Update outbound social impact and inbound risk assessments iteratively to help define scope.

- Evaluate options under consideration in terms of likely community and stakeholder acceptability. Identify potential fatal flaws and potential opportunities for enhancing 'social license to operate'.

During the Pre-Feasibility Study phase, the study team should:

- Carry out a detailed social and economic impact assessment in accordance with:
  - the study scope
  - regulatory ESIA requirements.
- Evaluate options under consideration in terms of public acceptability. Identify fatal flaws, mitigation measures and potential opportunities.
- Prepare detailed scope and cost of impact assessment work to be undertaken in the Feasibility Study. Update social risk and impact assessments iteratively to help define scope.
- Make land access agreements that include compensation provisions with landholders and users, as required in the local jurisdiction, or which will enhance social acceptance in line with advisable international good practice.
- Specify provisions for compensation for impacts that will be included in formal agreements with recognised land-connected Indigenous groups.
- Prepare a project affected area economic development strategy that includes options to support and develop:
  - local goods and services providers
  - local workforce.
- Establish a community concerns, complaints and grievance procedure for Feasibility phase.

In Feasibility, the study team should:

- Complete and approve ESIA studies required by regulators.
- Update social risk assessment and ESIA outputs iteratively to help finalise social impact mitigation measures.
- Finalise project affected area economic development plan.
- Establish site security and emergency/crisis management approach and protocols with relevant civic authorities for construction period and operations.
- Ensure a system to track and report on requirements of permits and approvals, consent agreements and proponent commitments is in place and implemented, with accountabilities assigned.
- Update the community concerns, complaints and grievance procedure for construction and operations.

Resettlement, economic displacement and in-migration

During the Scoping Study:

- Identify whether any physical resettlement, economic displacement and/or in-migration may result from any of the project options under consideration.
- In such cases, collate available baseline information on land tenure, land ownership, population, resource base (including formal and informal livelihood), housing and public assets.

- As required, prepare scope, schedule and budget for resettlement action plan (RAP), livelihood restoration plan (LRP) and/or in-migration management plan (including draft methodologies for determining livelihood compensation and in-migration management).

In the Pre-Feasibility Study phase, the study team should:

- Define the area and population subject to physical resettlement or economic displacement attributable to the project.
- Commence engagement with affected community groups about the resettlement and livelihood restoration planning process and in-migration possibilities. Prepare RAP, LRP and/or in-migration management plan as required (in accordance with legal requirements and consistent with recognised international standards).
- Start implementation of RAP and/or LRP if required by the project development schedule, and if not, subject to the final investment decision.

In Feasibility, the study team should:

- Start implementation of RAP and/or LRP if required by project development schedule, and if not, subject to the final investment decision.
- Implement the concerns, complaints and grievance procedure specific to RAP and LRP, including evaluation, monitoring and reporting plan.

Social closure planning

Social aspects of closure should be included in closure planning considerations set out in the earlier 'Closure planning' section.

Specifically, in the Scoping Study phase, the study team should:

- Consider social aspects and public use aspects of identified post-closure land uses and repurposing options.
- Identify which stakeholders would need to be consulted early in relevant closure-related discussions as part of ESIA and project permitting.

During the Pre-Feasibility Study phase, the study should:

- Describe options for post-closure land transfer that could affect the selection of preferred development pathways.
- Identify all specific social and economic parameters that will inform closure aspects of the ESIA, as required.
- Establish base-case and best-case scenarios for post-closure asset and liability transfer as part of the options analysis.
- Describe post-closure vision scenarios that have emerged out of affected community consultation.

In Feasibility:

- Describe preferences for post-closure land uses, repurposing options and post-closure asset and land transfer that have emerged out of affected community consultation.
- Describe local workforce transition, residual asset transfer and any relevant civic governance considerations in closure plans, including any plans required by regulators or others (eg lenders and insurers).

3

GEOLOGY AND MINERAL RESOURCES

INTRODUCTION

Elements of geological understanding and the Mineral Resource model are key foundation inputs that underpin the technical and economic assessment of a project at each phase.

At the Scoping Study phase, there is an order of magnitude technical and economic study of the potential viability of the Mineral Resources, which is insufficient to support an Ore Reserve. This level of study is typically an initial economic evaluation of the project, used internally by companies for projected, comparative and planning purposes. Appropriate assessments of realistically assumed Modifying Factors are used based on a combination of directly gathered project data, together with assumptions borrowed from similar deposits or operations to the current project.

At the Pre-Feasibility Study phase there is a comprehensive range of options for the technical and economic viability of a project available such that the project has advanced to a stage where a preferred mining method is established, and an effective method of mineral processing is determined. It includes a financial analysis based on reasonable assumptions on the Modifying Factors. It also includes evaluation of other relevant factors that are sufficient for a Competent Person, acting reasonably, to determine if all or part of the Mineral Resources may be converted to an Ore Reserve at the time of reporting.

A Feasibility Study is a comprehensive technical and economic study of the selected development option for a project. It includes appropriately detailed assessments of Modifying Factors along with other relevant operational factors and a detailed financial analysis that are necessary to demonstrate, at the time of reporting, that extraction is reasonably justified. The results of the study may form the basis for a final decision by a company or financial institution to proceed with, or finance, the development of the project.

As the Mineral Resource model is a key input for estimating Ore Reserves, it is essential that the input data and processes used to generate the model are robust, meet acceptable industry standards and meet the requirements for the level of confidence required for the various study phases. This chapter describes the key data required regarding a Mineral Resource for each level of study, along with the information and documentation to support the model as an input to the study.

The requirements and key geology and Mineral Resource considerations for each study phase are presented in Table 3.1. They are further described in this chapter.

TABLE 3.1

Geology and mineral resources study phase guide.

GEOLOGY AND MINERAL RESOURCES					
Activity	**Deliverable**	**Description**	**SS**	**PFS**	**FS**
Geology	Report, including maps and sections that describe the regional and deposit-scale geology.	• Regional geology – description of the metallogenic province, regional-scale lithology, geological age information and relevant stratigraphic information. • Deposit geology – description of the deposit geology including lithology, alteration, oxidation, structural controls and mineralisation.	✓	✓	✓
Exploration and Mineral Resource definition	Report with tables, maps and sections including details of work completed: • exploration overview • geological mapping • geochemistry and geophysics • drilling.	• Exploration overview – description of historical and recent exploration activities, summary of work completed, results and conclusions. Includes summary of any mining activities. • Geological mapping – description of work completed, results and conclusions. • Geochemistry and geophysics – description of work completed, results and conclusions. • Drilling – chronological description of all drilling completed and used to inform each study phase. Statistical analysis of different drill types and drill campaigns to quantify the quality of the drill data and identify any biases that may affect confidence. Tabulation of total number of holes and metres for the various drill methodologies. Key findings of any identified biases.	✓	✓	✓
Data inputs	Report detailing the items below and including key findings and conclusions: • logging and data acquisition • database and validation • survey data • sampling and assaying • QAQC • recovery data • bulk density • sample security and storage.	• Logging and data acquisition – description of the variables logged, logging interval and explanation of which variables are used to inform the Mineral Resource model. • Database and validation – description of database system, validation processes and independent reviews or audits. • Survey data – description of topographic, drill collar and downhole survey methods, frequency and confidence. • Sampling and assaying – description of sample collection and preparation process, assay techniques and confidence. • QAQC – description of QAQC type, source, frequency and confidence. • Recovery data – description of type of data and confidence. • Bulk density – description of methodology, source, quantity of data and how density is applied in the Mineral Resource model. • Sample security and storage – description of process.	✓	✓	✓
Mineral Resource estimation	Mineral Resource block model and supporting technical report detailing: • geological and domain modelling • data handling • estimation inputs, methodology and parameters • model validation • Mineral Resource classification • Mineral Resource reporting • RPEEE.	• Geological and domain modelling – description of the criteria used for domaining. • Data handling – description of data manipulation and handling. • Estimation inputs, methodology and parameters – description of all estimation-related inputs including top cutting, spatial continuity analysis, block model size, search and sample selection strategy. • Model validation – description of validation methods, key findings and confidence. • Mineral Resource classification – description of criteria used and confidence. • Mineral Resource reporting – table of Mineral Resource with description of any qualifying constraints. • RPEEE – description of criteria used and justification for constraints applied that represent the RPEEE applied to the Mineral Resource.	✓	✓	✓

TABLE 3.1 CONT...

Activity	Deliverable	Description	SS	PFS	FS
Geometallurgy	Report including description of data, key findings and conclusions. Geometallurgy models (depending on data collected) including: • mineralogy model • litho-geochemistry model • spatial model • mathematical model • ore control model • processing model – addressed in Chapter 5.	• Mineralogy model – prediction of mineralogy from geochemistry. • Litho-geochemistry model – prediction of material types. • Spatial model – 3D wireframes. • Mathematical model – classification and regression models to predict metallurgy response from rock properties. • Ore control model – adjustment of rock properties to maximise metallurgy response.		✓	✓
Exploration targets and potential	Report, including maps and figures that describe exploration potential and prospectivity.	Information to demonstrate the future exploration prospectivity for the project, supported by proposed resource delineation and exploration drill programs, and proposed budget and timing.		✓	✓
Technical risks and opportunities	Report that includes a ranked list of identified technical risks (related to geology and the Mineral Resource) and opportunities.	A list of identified key technical risks and opportunities, ranked in order of materiality and including: a description of the issue, the nature and materiality of the risk/opportunity, and recommended action to mitigate the risk or enable the opportunity.	✓	✓	✓

SS = Scoping Study; PFS = Pre-Feasibility Study; FS = Feasibility Study; QAQC = quality assurance and quality control; RPEEE = reasonable prospects for eventual economic extraction.

GEOLOGY

Regional geology

The metallogenic province in which a project is located is, in most cases, already well researched and documented from sources such as government mapping, previous company work in the area, as well as research projects including honours, master and PhD thesis work. The understanding of the regional geology is generally already well established from the Scoping Study phase. The metallogenic province is typically a key driver for exploration activities in the early exploration phase of the project.

Data required to inform and support the study includes a description of the metallogenic province that hosts the project, regional-scale lithology, geological age information and relevant stratigraphic information, along with relevant geological maps to support the description.

Deposit geology

The understanding of the deposit geology generally increases as each study level progresses due to the increased amount of information available from exploration and Mineral Resource delineation activities that form part of the technical and economic assessment. Data sources include geological mapping, geophysical and geochemical surveys, and drilling.

Data required to inform and support the study includes a detailed description of the deposit geology including descriptions of lithology, alteration, oxidation, structural controls and mineralisation. Representative geological maps, cross-sections and long sections are required to support the description.

It is expected that the understanding of controls on mineralisation, grade distribution and mineralogy will be better understood at a Pre-Feasibility Study and Feasibility Study phase than at a Scoping Study phase. Activities completed to support this understanding include detailed data analysis, petrology and mineralogy studies, sample heterogeneity test work and metallurgical test work.

EXPLORATION AND MINERAL RESOURCE DEFINITION

Exploration overview

Most projects will have undergone a lengthy period of various exploration activities conducted by different companies over many years before they reach a Scoping, Pre-Feasibility or Feasibility Study phase.

Data required to inform and support the study includes a chronological description of all historical and recent exploration activities completed. This should also include a summary of any mining activities completed. Information should include the company name, dates when work was undertaken, along with a summary of the work completed, results and conclusions. Summarising this data in a table form is usually the most efficient method of presenting the data. The reporting should clearly separate historical work from work completed by the current company that owns the project. The data should be supported by maps and figures that clearly illustrate the nature and extent of the work. For example, for drilling, a map showing all drill collars coloured by year, drilling type or campaign is a simple and effective way to visually present the data.

Geological mapping

Geological mapping provides one aspect of support for the understanding of the geology of the project. The availability and extent of mapping depends on the geological and topographic terrain. Where mapping can be undertaken, this is usually completed at an early to late exploration phase and so is already available as support data from scoping.

Supporting data should include geology reports and geological maps completed for the project that detail the extent of work undertaken, results and conclusions.

Geochemistry and geophysics

Geochemical and geophysical survey data provides an exploration targeting tool usually undertaken as part of early to advanced exploration phases of the project. Analysis and results from this data feed into the drill targeting strategy as a project progresses through the various phases. More detailed and spatially focused surveys may be completed at the advanced project phases to better understand the deposit. The types of geochemical and geophysical surveys used depend on the geography, geology and style of mineralisation, which will determine the most appropriate methods and sample or line spacing to be used. Geochemical surveys may include rock geochemistry, soil geochemistry and stream sediment sampling. Geophysical surveys may include ground and aerial surveys for magnetics, gravity, induced polarisation, resistivity, as well as use of satellite imagery, along with downhole surveys as typically used in coal and becoming more common in other deposit types.

Supporting data should include reports, plan and cross-section figures completed for the project that detail the extent of work undertaken, results and conclusions.

These investigations are typically undertaken in exploration or for the Scoping Study phase.

Drilling

Drill data is the most significant information that is used as an input for generating a Mineral Resource model, which in turn is used to generate an Ore Reserve estimate and forms the foundation for technical and economic analysis of the project at all phases. The drill data provides geological, geotechnical, spatial and analytical information that is fundamental to applying a confidence assessment by the Competent Person. For this reason, it is critical that robust quality assurance and quality control (QAQC) and validation processes are in place to quantify uncertainty associated with the drill data used for modelling.

The quantity and spacing of drilling typically increase through the study phases, as an increase in confidence is required moving from Scoping Study through to Pre-Feasibility. The relevant spacing and confidence depends on the style of mineralisation and proposed mining method. For example, a porphyry copper deposit with low grade variability and proposed open pit mining method will most likely have a wider spaced drill pattern that supports the various Mineral Resource categories, than say a narrow vein gold deposit with high grade variability and a proposed underground mining method requiring closer spaced drilling.

As the project moves to higher level studies there is a requirement for drilling focused on the technical areas that support the Modifying Factors including geotechnical, metallurgical, hydrological, planned infrastructure and environmental. Ideally some holes will have been drilled to support these areas in a Scoping Study, although often Mineral Resource delineation holes are multipurposed to obtain this data at this stage.

During the Pre-Feasibility and Feasibility Study phases, strategic standalone holes are required to collect this data to support the study. The number and location of holes required to provide this data will depend on the deposit style, mineralisation style, structural setting, scale of the deposit and proposed mining method and infrastructure.

Data required to inform and support the study includes a chronological description of all drilling completed and used to inform each study phase. Statistical analysis of different drill types, drill campaigns and sampling or assaying protocols should be completed and documented to quantify the quality of the drill data and identify any biases that may affect confidence. Information should be summarised in a table and include the total number of holes and metres for the various drill methodologies. Key findings of any identified biases should be included in the study report. The data should be supported by maps and figures that clearly illustrate the nature and extent of the work.

DATA INPUTS

The data inputs to the Mineral Resource to represent the informing variables used to estimate the Mineral Resource, as well as provide a basis for assigning confidence in the Mineral Resource classification, which feeds into the Ore Reserve process. The following topics cover the key areas and variables that must be considered and reported at each study phase.

Logging and data acquisition

The logging system is typically developed during exploration and needs to be flexible to incorporate new and additional data, such as detailed geotechnical logging, which is required as the project moves forward into higher study levels.

Data logging and data acquisition starts during the exploration phase of a project and continues throughout all study phases. Key variables that need to be captured at all study phases include lithology, alteration, oxidation, mineralisation, structure, core recovery, hardness and basic geotechnical data.

A description of the variables logged and logging interval should be documented to support the study, along with an explanation of which variables are used to inform the Mineral Resource model.

Database and validation

In most cases the database system is already well established during the exploration phase of the project and should not be affected by the subsequent study phase. The database system selected should have the capacity to easily integrate additional data required to support the study as the project moves forward, for example metallurgical and geotechnical data.

In particular, the database should have robust validation protocols ensuring that input data is validated against compliant and valid codes and data ranges, and that the validation analysis can be output so that project personnel can identify and assess the suitability of data for use in the Mineral Resource estimate. A description of the type of database system, data entry and validation processes should be documented to support the study, along with documentation of independent checks, reviews or audits on the database that will support confidence in the data used in the Mineral Resource estimate.

Survey data

Survey data, both surface and downhole, is critical in assigning confidence in the spatial location of the deposit area and the Mineral Resource model. Data required includes current topographic survey data, drill collar survey data and downhole survey data. Typically, drill collars are sourced from registered survey technicians, and downhole surveys from appropriate downhole measuring devices. It is important to note that deposits hosted in or influenced by magnetic rocks require suitable survey methodologies that account for the magnetic influence of the rocks.

Topography will be sourced from light detection and ranging (LIDAR) survey (or equivalent) and is often only available in a limited amount in the Pre-Feasibility Study phase. For the Feasibility Study phase, it is expected that the area being studied is covered.

All study phases should be supported by valid industry standard survey methods that provide a high degree of confidence in the spatial location of the data used to support the Mineral Resource. A description of the type of survey method and frequency of downhole surveys should be documented to support the study. Data excluded from the Mineral Resource due to poor confidence in survey data should be documented.

Sampling and assaying

Sample data is a critical input for estimating and assigning grade into the Mineral Resource model. Sampling interval, preparation procedures and assay techniques must be of an industry accepted standard and appropriate for the type of deposit and style of mineralisation.

Elements selected for assay should be selected to not only support the Mineral Resource estimation but to also consider waste mineralogy and potentially deleterious elements that may impact metallurgical recovery and/or environmental factors that need to be addressed.

Typically, a wide suite of elements is assayed during the exploration phase to better understand the geology and controls on mineralisation. However, often once a Mineral Resource is defined, the elements assayed are generally significantly reduced, with a focus on the economic components. It is important that as the project progresses through the study phases there is effective communication between all technical disciplines to ensure that all required assay data is collected that will feed into the various aspects of the study at each phase.

Additional detailed assaying such as heterogeneity test work and size fraction analysis is usually done during the Pre-Feasibility Study and Feasibility Study phase as this information feeds into the metallurgy studies, design of the processing plant and planning for optimal production sampling when mining commences.

All study phases should be supported by high-quality sample collection, preparation and assay procedures. A description of the sample collection, preparation and assay techniques should be documented to support the study. Data excluded from the Mineral Resource due to poor confidence in quality or identified bias should be documented.

Quality assurance and quality control

QAQC for the sample data used in the Mineral Resource model is essential for monitoring precision, accuracy and bias of assays used to estimate grade in the model and assign confidence in the data. The type and frequency of QAQC data should meet industry accepted standards and be appropriate for the type of deposit style of mineralisation and grade thresholds. Typically, QAQC protocols will have been established during exploration and should be maintained at the same levels through all the study phases. Ideally this will include heterogeneity assessments to support the selected sample preparation protocols. Types of QAQC samples include regular inclusion of certified reference material, field and laboratory duplicates, coarse reject and pulp duplicates, blanks and umpire samples. It is also important that the QAQC results are monitored on a regular basis and any identified discrepancies or non-compliance followed up immediately.

All study phases should be supported by high-quality and appropriate QAQC protocols that are regularly monitored. A description of the QAQC procedure, type, source, frequency, preparation and assay techniques should be documented to support the study, as well as reports detailing results and key findings. Data excluded from the Mineral Resource due to poor confidence in quality or identified bias should be documented.

Recovery data

Sample recovery data for drill samples should be recorded to assist in assigning confidence to the sample quality of the data informing the Mineral Resource. In the case of air drilling, such as percussion or reverse circulation, the weight of the sample at the sample collection point should be recorded. The amount of moisture should also be recorded as the quality of wet samples may be negatively compromised due to the impact of water smearing grade within the drill hole. For diamond drill core, the amount of core recovered per drill run should be recorded and zones of core loss accounted for in the resource estimation process.

Supporting data should include documentation, and relevant graphs and figures that detail the key findings and conclusions. Data excluded from the Mineral Resource due to poor confidence in quality or identified bias should be documented.

Bulk density

Bulk density data is applied and used to estimate and report tonnages of the Mineral Resource. It is important to use industry accepted practices to measure sample density and obtain a robust quantity of values that represent different lithologies, mineralisation styles and oxidation types for the project. During the exploration phase, densities may be assumed and applied as a generic value based on lithology or oxidation type. However, once a project progresses to a Scoping Study or higher study phase it is necessary to obtain representative density data that reflects the main material types of the deposit. There are various methods available to obtain sample density data including the Archimedes method on core samples, and downhole density logging as used in coal deposits. If the Archimedes method is used, porous material should be accounted for by using a coating, such as wax or plastic wrap, to protect the core before water immersion. If the material has a high clay content, then it should be dried before weighing. Sample density results may not be directly representative of the bulk density of the larger rock mass due to the non-representative selection of the samples for testing, the testing method or the physical characteristics of the bulk rock mass. Sample densities therefore may require modification to represent the bulk rock mass density and this requires due consideration and discussion.

A description of the density methodology, source, and quantity of data for different material types as well as the conversion of sample to bulk densities should be documented to support the study, as well as a description of how density has been applied in the Mineral Resource model.

Sample security and storage

It is important to be able to demonstrate that the chain of custody of samples used to inform the Mineral Resource is secure and can be accounted for at all stages of the sample handling process. Systems are usually set up during exploration and continue through all study phases. Areas to be addressed include:

- secure handling of samples at the drill rig
- transportation to the sample logging and storage facility
- sample handling and storage at the storage facility
- despatch and transport of samples to the laboratory
- process and accountability of receiving samples at the laboratory
- storage at the laboratory
- sample preparation and assaying at the laboratory
- process for receiving assay data back and integrating into the database.

Supporting data should include documentation describing the process and steps taken to mitigate potential risk associated with the security of the samples that could compromise the integrity of the data used in the Mineral Resource.

MINERAL RESOURCE ESTIMATION

The Mineral Resource model must be fit for purpose and is required as an input for all study levels. The confidence in the classification of the Mineral Resource will usually increase as more drilling is completed moving from a Scoping Study to the Pre-Feasibility Study and Feasibility Study phase.

At the Scoping Study phase there will not have been an Ore Reserve established and the Scoping Study may be relying on lower confidence Inferred Mineral Resource material to inform the study. Data to support the Modifying Factors may not yet be available for the project and may be applied based on assumptions from similar deposits.

As the project moves into a Pre-Feasibility Study and Feasibility Study phase there is a requirement that the Mineral Resource will have, at least, sufficient Indicated Mineral Resource material to support an Ore Reserve and a technically achievable and economically viable project, and that direct gathered project data has been acquired to support the Modifying Factors.

The following topics describe the key areas of the Mineral Resource model process that must be considered and reported to support a study.

Geological and domain modelling

Geological and domain modelling is the first step in defining the criteria that will control the grade estimation in the Mineral Resource model. Understanding the controls on mineralisation and defining appropriate domains that reflect this are essential steps in the modelling process. All subsequent steps in the estimation process depend on the appropriateness of the domaining strategy, and the domaining strategy applied is dependent on deposit and mineralisation style.

A description and justification of the criteria used for domaining along with relevant statistical analysis, figures, plans and sections should be documented to support the study.

Data handling

Manipulation and treatment of data includes:

- consideration of missing data
- consideration of data below assay detection limits
- determination of composite length
- determination of top cutting strategy
- exclusion of data due to lack of confidence or other reasons.

Support data for all study phases should include documentation including a description and justification for all aspects associated with the data handling that is used in the Mineral Resource.

Estimation inputs, methodology and parameters

Statistical analysis of domains, top cutting strategy, spatial continuity analysis and qualitative kriging analysis (if appropriate) are the sequential steps undertaken to establish the various inputs and parameters that control grade estimation in the Mineral Resource model. These parameters (if applicable) include:

- model parent block size
- subcelling
- search parameters
- variogram parameters
- minimum and maximum samples
- drill hole restriction
- dynamic anisotropy
- high grade restriction
- discretisation
- selective mining unit adjustment.

Estimation methodology should be selected based on appropriateness for the deposit style and data support.

A description and justification of all estimation parameters applied should be documented to support the study.

Model validation

Validation of the model should be completed to provide support for the block model grade estimates. Procedures for validation include:

- visual checks of block grade estimates against input drill data
- statistical analysis of block grades against composite grades
- swath plots of average grade comparison between composite data and block grades in northing, easting and elevation profiles
- cross-validation estimates or alternative estimation methodology results.

Where appropriate, for example where selective mining is expected and a robust grade–tonnage relationship is required, the validation may include analysis of the information effect, change of support and related checks on the expected selectivity and a discussion on the expected confidence in the presented grade–tonnage relationship.

All study phases should be supported by block model validation against the input data to demonstrate the robustness of the grade estimation. Support data for a study should include a description of the validation methods completed, accompanied by tabulated statistics, relevant figures and key findings.

Resource classification

Mineral Resource classification (Measured, Indicated and Inferred) reflects the level of confidence in the Mineral Resource and is agreed to and signed off by the relevant Competent Person. The confidence in the classification of the Mineral Resource will usually increase as more drilling is completed moving from a Scoping Study to Pre-Feasibility Study and Feasibility Study phases. At the Scoping Study phase Inferred Resource material can be considered, with no requirement for Indicated or Measured Mineral Resources. Once a project moves to Pre-Feasibility Study and Feasibility Study phase it is required that the Mineral Resource has, at least, Indicated Resources that will convert to Probable Ore Reserve, and that direct gathered project data has been acquired to support the Modifying Factors applied in the Ore Reserve.

Parameters to be considered in the Mineral Resource classification include:

- the distribution and density of drill data
- confidence in interpreted geological continuity of the mineralised zones
- confidence in sample and assay data
- confidence in the resource block estimates.

Support for the study should include documentation of the criteria used and justification for the Mineral Resource classification applied and its relative accuracy and confidence, including whether this relates to local or global estimates, along with figures showing the relevance of classification to data support.

Resource reporting

Mineral Resource reporting should state tonnage and grade for the contributing economic commodities to the project as well as whether reported inclusive or additional to the Ore Reserve. The content of information required to meet the Code for Reporting of Exploration Results, Mineral Resources and Ore Reserves (the 'JORC Code') reporting guidelines is the same irrespective of the study level. Constraints applied to the Mineral Resource for reporting, such as cut-off grade, economic mining constraints (eg open pit or underground stope development), and any current mine depletion, should be noted in the report table.

The study should be supported by a table of the Mineral Resource broken down by classification at a stated cut-off grade or equivalent value with explanation of the derivative equivalent value if applicable.

Reasonable prospects for eventual economic extraction

Consideration must be given to the viability of reasonable prospects for eventual economic extraction (RPEEE) with respect to Mineral Resources for all study phases. An optimised pit shell or proposed underground development, at an assumed commodity price, is typically applied to the model to report a Mineral Resource that meets JORC Code reporting guidelines. The confidence in the RPEEE criteria will generally increase as the project progresses through the study pipeline.

The study should be supported by documentation of a description and justification for the constraints applied that represent the RPEEE criteria applied to the Mineral Resource.

GEOMETALLURGY

Combining geology and metallurgy studies with data analysis and mathematical modelling is a powerful way of accounting for metallurgy Modifying Factors in resource estimation. Data analysis is used to identify relationships between geology and metallurgy that can be exploited by mathematical models for identifying metallurgical domains and predicting metallurgy response. The process requires comprehensive geology and metallurgy data that is representative of the entire Mineral Resource, which is typically not available, even in the Feasibility Study phase.

This is addressed further in Chapter 5, Metallurgy.

Elements related to geology are addressed here.

Data requirements

It is important to consider geometallurgy at the beginning of a project so that all necessary and sufficient data is collected. The progression from Scoping Study to Pre-Feasibility Study and Feasibility Study should not alter what data is collected. However, by the nature of the focus of the project phases, data collection may vary through the study phases.

Data that may be collected includes:

- trace element geochemistry to characterise protolith under varying degrees of alteration and oxidation
- whole rock geochemistry for characterising gangue mineralogy
- rock hardness
- rock abrasiveness
- density
- geology logging of lithology, alteration, oxidation, mineralisation type and texture
- geological interpretation of rock domains and mineralisation
- mineralogy, which can be combined with whole rock geochemistry, to identify gangue and minerals of economic interest
- metallurgical test work (see Chapter 5).

Geometallurgy modelling studies

The geometallurgy model may comprise any and all of the following: a mineralogy model, litho-geochemistry model, mathematical models, spatial model, ore control model and process model. These models are typically used to add additional attributes to the resource block model, which can then be used to evaluate the Mineral Resource with respect to the metallurgy Modifying Factors. However, it is not unusual for these models to commence development as late as a Feasibility Study phase.

Mineralogy modelling

Analysis of rock types to determine gangue and minerals of economic interest, mineralogy, texture and mineral associations and its likely effects on metallurgy. Data required includes petrology and mineralogy data. This work can be undertaken during the early phases of exploration and Mineral Resource definition. Mathematical modelling can be performed to allow prediction of mineral species from the geochemistry.

Litho-geochemistry modelling

Analysis of the geological logging, trace element geochemistry and whole rock geochemistry to determine broad litho-geochemistry domains for geological interpretation, resource modelling, and collection of samples for initial metallurgy test work starts in the Scoping Study phase. This study is generally conducted before any metallurgy sampling and test work has been done. The litho-geochemistry model is then used to assist with identifying material types (in conjunction with mathematical modelling), which are then used for selecting metallurgy samples.

Spatial modelling

The drill hole data is encoded with material type predicted using the geometallurgy classification model and from the litho-geochemistry model. Geological 3D spatial interpretation of the material types is then performed to define spatial geometallurgy domains for Mineral Resource modelling. Average metallurgy response can then be assigned to each geometallurgy domain for further mining and economic studies. Issues with spatial continuity of the identified material types can be fed back into improving the geometallurgy mathematical modelling of material types.

Mathematical modelling

Mathematical modelling incorporates the data analysis of the geology and metallurgy data. The mathematical modelling can be used to support other geometallurgy modelling studies such as mineralogy and litho-geochemistry modelling, particularly when there is no metallurgy data. Machine learning can be used to build mathematical models for predicting mineralogy given geological domains (lithology, oxidation, etc) and geochemistry, identifying and predicting material types, and for optimal sampling for metallurgy test work. Outputs from the modelling can be used to assist geological interpretation and domaining for Mineral Resource estimation and add additional attributes to the geological block model.

Ore control modelling

Statistical analysis and mathematical modelling tend to focus on the correlation between independent variables and dependent variables, however, the independent variables may not directly cause or affect metallurgy response. Geology rock properties are assumed to have some causal relationship with metallurgy, so most geology variables will affect metallurgy to varying degrees. Ore control modelling is more interested in adjusting for different rock properties to maximise the metallurgy response.

EXPLORATION TARGETS AND POTENTIAL

In assessing the economic viability of a project at the various study phases, it is also necessary to understand the potential for future growth of the project and the projected impact on the life-of-mine. Items to be considered include:

- an assessment of the likelihood and potential scale of expansion of the existing Mineral Resource
- estimated percentage of conversion of Mineral Resource to Ore Reserve with planned resource delineation drilling
- potential for additional Mineral Resource sources through exploration activities.

Information should be provided to demonstrate the future exploration prospectivity for the project. This should include documentation and figures to describe and illustrate proposed resource delineation and exploration drill programs, supported by proposed budget and timing.

TECHNICAL RISKS AND OPPORTUNITIES

A list of identified key technical risks related to geology, the Mineral Resource data inputs and the Mineral Resource model should be documented for each study phase. Identified risks will typically already have been discussed in the relevant report sections of the topics presented in this chapter. However, it is helpful to summarise them all in a separate section for clarity. The risks should be ranked in order of materiality and include: a description of the issue, the nature and materiality of the risk, and recommended action to mitigate the risk.

Similarly, a list of identified key opportunities related to the geology and Mineral Resource should be documented and summarised. The opportunities should be ranked in order of materiality and include: a description of the issue, the nature and materiality of the opportunity, and recommended action to enable the opportunity.

ACKNOWLEDGEMENTS

The author would like to acknowledge the valuable contribution by my ERM Australia colleague Matthew Nimmo, who compiled the section on geometallurgy. The author would also like to acknowledge the support provided by the ERM Australia management team for allowing time to write the chapter, provide approval for publication and also the helpful peer review by work colleagues.

4

MINING AND ORE RESERVES

INTRODUCTION

The minimum outputs of a study required for assessing the mining aspects of a mineral project include:

- selecting a mining method
- developing a mine design
- defining ore and waste
- estimating ore and waste quantities
- scheduling ore and waste quantities
- determining technical and economic viability
- estimating, classifying and reporting an Ore Reserve under the guidelines of the Code for the Reporting of Exploration Results, Mineral Resources and Ore Reserves (the 'JORC Code') (JORC, 2012).

A Scoping Study does not identify the preferred mining configuration as it is focused on demonstrating reasonable prospects for technical and economic viability of a mineral project that justifies proceeding to a Pre-Feasibility Study and provides the inputs to the Pre-Feasibility Study. A Pre-Feasibility Study assesses the mine development options to identify the preferred option and economic case for proceeding to a Feasibility Study and provides the inputs to the Feasibility Study. The Feasibility Study then assesses that preferred configuration in detail.

Often an objective of undertaking studies on a mineral asset is to support an estimate of a Mineral Resource and Ore Reserve, which might then be publicly reported to fully inform potential investors about the mineral asset and to assist in improving the technical basis for further development of the mineral asset. The requirements for estimating and reporting an Ore Reserve must be understood from the start of each study phase to ensure that the results of the study meet the requirements for an Ore Reserve. For example, a Scoping Study is not considered by the JORC Code to be a reasonable basis for an Ore Reserve, which requires a Pre-Feasibility Study or a Feasibility Study.

In this chapter, 'Ore Reserve', as defined by the JORC Code, is stated as 'Reserve', and 'Mineral Resource' as 'Resource', as most of the principles are applicable to other codes, which may use the term 'Mineral Reserve' and commodities such as coal. The term 'ore' is used generically, and may refer to base or precious metals, coal, diamonds, industrial materials, bulk materials, evaporites or brines, or any material containing minerals of economic interest.

The study phase guide

The study phase guide table (Table 4.1) provides a summary of material issues that need to be defined and assessed in a study or an operating mine's plan at the required level of detail and confidence to determine the mining requirements for the mineral project. This also supports development of a mine plan appropriate to the deposit and to the level of accuracy required by the type of study being undertaken.

TABLE 4.1

Mining and Ore Reserves study phase guide.

MINING AND ORE RESERVES					
Activity	**Deliverable**	**Description**	**SS**	**PFS**	**FS**
Establish mine planning criteria	Basis of design	Spreadsheet or document, managed by document control, and updated as required.	✓	✓	✓
Mine scale	Production rate(s) Mining rate(s)	Ore processing and/or material handling rates considered. Material movement, and equipment and personnel requirements.	✓	✓	✓
Selectivity and cut-off value/grade analysis	Definition of ore	Trade-off between size of selective mining unit and quality and determining the material mined and processed or considered as waste.	✓	✓	✓
Processing method selection	Method(s) of processing considered	Description of process flow sheets, processing plant equipment and infrastructure and calculation of throughput, recovery/yield, operating cost and capital cost.	✓	✓	✓
Ground and surface water management	Water management plan	In a form suitable for estimating cost and determining whether water use and water capture can be balanced, and appropriate environmental and social outcomes can be achieved.	✓	✓	✓
Ground support and geotechnical stability	Geotechnical design parameters	In a form suitable for developing a geotechnically robust and safe mine design, with subsidence controlled as required (caving or goaf).	✓	✓	✓
Mining method selection	Method(s) of mining considered	Description of and justification for mining methods considered, and method-specific infrastructure required.	✓	✓	✓
Mine design based on rock mass characteristics and Resource	Design(s) specific to the characteristics of the deposit being studied	A spatially correct design showing economic limits of extraction, means of access for ore, waste, personnel, materials and equipment and demonstrated alignment with geotechnical recommendations.	✓	✓	✓
Option analysis	Options for marketing, mine strategy, infrastructure and processing	Options to be studied in a Pre-Feasibility Study to identify the preferred project development pathway.	✓	✓	
Mining equipment selection	Equipment fleet specific to the deposit being studied	Description, size and number of equipment units required for the rock mass characteristics and mine design and number of working areas.		✓	✓
Mining infrastructure and services	Infrastructure and services required to mine, process and transport products to market	In a form suitable for ensuring sufficient power, water, and other services are available, estimating a cost and determining whether appropriate environmental and social outcomes can be achieved.	✓	✓	✓
Waste management	Plan and cost estimate to manage waste material	In a form suitable for estimating a cost and determining whether appropriate environmental and social outcomes can be achieved.	✓	✓	✓
Interaction with tailings disposal and backfill	Tailings and backfill management plan	In a form suitable for estimating a cost and determining whether appropriate environmental and social outcomes can be achieved.	✓	✓	✓

TABLE 4.1 CONT...

Activity	Deliverable	Description	SS	PFS	FS
Mining schedule	Mining, processing and product realisation schedule	Schedule of physical quantities and qualities in a form suitable to estimate mining revenue and costs for economic evaluation.	✓	✓	✓
Mining revenue and cost estimation	Owner/contractor trade-off, organisation charts, economic evaluation	Revenue generated from mining products and costs of generating the product to the point of sale, including mine closure costs.	✓	✓	✓
Mine closure planning	Conceptual plan and cost estimate for mine closure	In a form suitable for estimating a cost and determining whether appropriate environmental and social outcomes can be achieved.	✓	✓	✓
Modifying Factor application to the mine plan	Conversion of Resource to Reserve	The Reserve table, classified into appropriate confidence level of Proved and Probable Reserve, with notes critical to understanding the estimate.	N/A	✓	✓

SS = Scoping Study; PFS = Pre-Feasibility Study; FS = Feasibility Study.

Establish mine planning criteria

The JORC Code provides the general relationship between Exploration Results, Mineral Resources and Ore Reserves in Figure 4.1, which also sets out the considerations used to convert a Mineral Resource to an Ore Reserve, also referred to as the Modifying Factors (Figure 4.1).

The basis of design for a mining study needs to consider a wide range of factors, a guide to which are the material Modifying Factors shown in the JORC Code, and that are required to convert a Resource to a Reserve. Some of these Modifying Factors may be known before the start of the study, some may be defined by analysis during the study, and some may change as a result of the analysis in the study. The study is the basis for supporting the value and application of the Modifying Factors.

These considerations will include the key technical; economic; and environmental, social and governance (ESG) parameters and assumptions that will be used as the basis for analysis in the study:

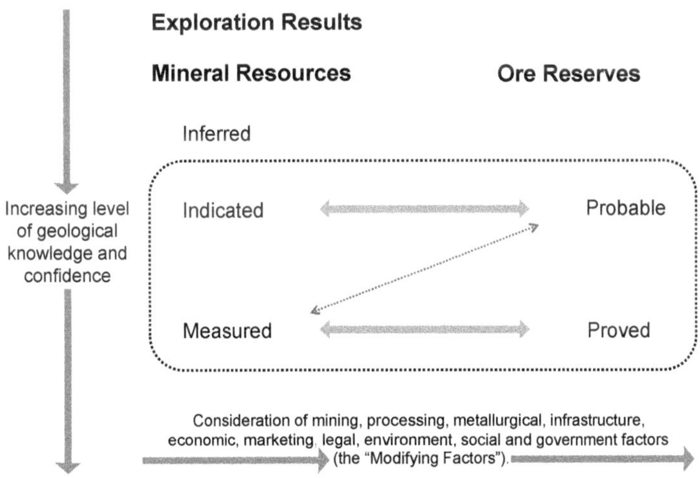

FIG 4.1 – General relationship between Exploration Results, Mineral Resources and Ore Reserves. Source: Figure 1 (Clause 12) of the 2012 JORC Code.

- *mining* – such as mining method, selectivity, dilution, and mining recovery, mine design parameters, method of quality control, geotechnical parameters, hydrogeology, mining hazards and control measures, production rates and sequences
- *processing* – such as processing method, throughput, recovery/yields, material hardness, product size and quality specification, contaminant limits and by-products
- *infrastructure* – such as source and availability of sufficient power, water, gas, and telecommunications, transportation methods for ore to processing and product to market, tailings management, accommodation and access roads
- *revenue* – such as product demand forecasts, product prices, exchange rates, royalties, penalty limits, treatment and refining costs, product specification and payability
- *costs* – such as initial capital costs, sustaining capital costs, and operating costs for mining, processing, infrastructure, product transport, and marketing and product realisation costs
- *ESG* – such as environmental, community, historical and cultural restrictions, legal requirements, noise, dust and vibration limits, as well as environmental licences and monitoring, tenure and government approvals.

It is not possible to list all of the considerations, as it will be different for each commodity (coal, base metals, precious metals or bulk commodity), mining method (open pit or underground, haul trucks or conveyors, shovels or dredges or draglines), location (remote or near communities, arid location or tropical), and operations model (owner or contractor, residential or fly-in, fly-out). However, the study group and/or Competent Person will be required to ensure all relevant considerations are included.

Less conventional mining techniques are likely to have a very different set of considerations (extraction of brines or evaporites, and deep-sea mining).

Importantly, the treatment of Inferred Resources in the mine plan is a key mine planning criterion for a Reserve to be an output of a Pre-Feasibility Study or Feasibility Study. Although there is no pathway for conversion of Inferred Resources to a Reserve (see Figure 4.1), the JORC Code does allow material derived from Inferred Resource to contribute to a Reserve as dilution and to inform the mine design (which is not allowed under some other mineral reporting codes), but not the economic evaluation. A study can include Inferred Resources in the design but Inferred Resources cannot be publicly reported or used to support the value of a project.

The key input assumptions for a Scoping Study might be derived from other similar projects, inputs for a Pre-Feasibility Study will be from the outcome of a Scoping Study and the outcome from the Pre-Feasibility Study will provide key input assumptions for a Feasibility Study. As technical assessment progresses in each study phase, there are likely to be fewer assumptions and more defined parameters. For example, a Pre-Feasibility Study might have more assumptions than a Feasibility Study, as a Feasibility Study will undertake additional analysis to convert Pre-Feasibility assumptions into defined Feasibility Study parameters.

Mine scale

A key decision for all mining operations is to define the scale of the operation in terms of, for example, the quantity of ore processed, product generated or material mined. To determine the optimum or preferred scale of the mine, the dimensions and geometry of the mineralisation, its location, the marketing of the products from the mine, environmental and community issues, the capital cost of construction, unit operating costs, and the expected or desired return on investment are key aspects that must all be considered.

The defined production rate will assist in determining issues such as the method of mining, method of processing, the type and size of mining equipment and personnel required, material handling and logistics facilities required, operating costs, and the capital cost for the construction of facilities. It will also have a flow-on effect to issues such as the rate of mine dewatering, ventilation requirements and stope backfill logistics.

There is no universal rule as to how to define the scale of the operation. The scale of an operation may be defined by any of, or a combination of:

- the amount of product that can be transported or sold into a limited market
- the amount of ore that can be processed (or processed through a particular part), or product that can be produced by an existing processing plant
- the total amount of ore or waste that can be mined by a particular mining method (for example hoisting limits underground or trucking limits in an open pit) or type of equipment (dredge)
- the limits to the rate of advance of the mining operation, such as development rate in an underground mine or bench turnover rate in an open pit
- legislated or permitted limits due to ESG or other issues.

A Scoping Study is likely to assess only one production rate to determine potential technical and economic viability. A Pre-Feasibility Study may consider more than one rate to identify the optimum or preferred technically feasible economic production rate. There are optimising evaluation techniques that use a 'Hill of Value™' approach of varying production rate along with a cut-off value to identify the optimum combination of production rate, cut-off value and stockpiling strategy that produces the highest economic value or the lowest risk to economic value.

Any one of the production rates considered in the Pre-Feasibility Study could form the basis of the Reserve, provided it meets the criteria of a Reserve (see the JORC Code). However, only one of the production rates assessed can be the basis for the Reserve.

A Feasibility Study would further evaluate and refine the preferred production rate identified as the preferred option from the Pre-Feasibility Study.

Selectivity and cut-off value/grade analysis

The inventory derived from the mine plan and used for mining scheduling (and the Reserve estimate) needs to be estimated at a specific cut-off value that is used to define ore and below which material is considered as waste. Cut-off values may be specific to a lithology or weathering unit, which may have different processing costs and/or different metallurgical recoveries.

The strategy and the criteria needed to define ore and waste need to be established when mine planning criteria and the scale of the mine have been established. The definition of ore is different for different commodities and can be based on:

- The quality required to generate a marketable product (typical of bulk commodities).
- What is required to make a profit when processed and sold (typical of base and precious metals):
  - when the processing plant is scheduled to be full, and each new block of ore processed displaces other ore and fixed costs are likely to be included
  - when the processing plant is under capacity, and each new block of ore processed does not displace other ore and fixed costs are likely to be excluded.

The definition of ore may be in terms of the grade/quality (such as g/t Au, per cent Cu, per cent Fe_2O_3) or a specific quality (such as calorific value or percentage ash) of the main commodity being mined. For polymetallic deposits, a metal equivalent (a single equivalent grade of one major metal), net smelter return (revenue derived from the products less realisation costs), or expected value (revenue – cost, where expected value above $0 is reported as ore) may be used. The metal prices and metallurgical recoveries used to calculate the values should be stated when reporting using equivalent grade or net smelter return.

For an open pit, the smallest parcel of ore used in mine planning to define a mining block is often referred to as the selective mining unit (SMU), which is the smallest block of continuous mineralisation that can be defined, and a loading unit can productively and selectively mine. There is link between the size of a SMU and orebody geometry, orebody knowledge, loading unit cutting tool (such as bucket width) size, and bench height.

For large-scale mines, large mining equipment is typically used and the SMU is proportionately larger. This commonly results in additional dilution and a lower mined grade, but as operating costs are lower it may also result in a lower cut-off value. A deposit therefore may have a different cut-off if it is mined as a selective small-scale mining operation by smaller equipment or a large-scale bulk mining operation with larger equipment.

For all levels of study, the cut-off strategy must be defined and the cut-off value or values need to be calculated. A mine plan (and the documentation for a Reserve) needs to have an accompanying cut-off value and explanation of how the cut-off value was calculated.

Processing method selection

A mine plan needs to be based on a specific method of processing the ore to extract the valuable products. If estimating a Reserve, the JORC Code states that 'an effective method of processing is determined'. Selection of a preferred, effective method of extraction of valuable minerals should consider specific, measurable performance characteristics of the deposit established during testing of available processing methods.

The selected processing method must take account of the commodity and type of mineralisation, and may be as simple as mining material and transporting it to a market for sale (directly saleable ore) or processing ore through a series of complex chemical reactions to make multiple products, each with strict product quality specifications. However, selection of a method of processing is critical to estimating what valuable products can be generated from the ore, what qualities those products will have, how much of each product will be generated and what revenue will be generated from its sale. Without which, it is not possible to estimate revenue, capital cost of construction or operating cost.

When the range of production rates to be assessed has been defined, consideration must be given to the methods of processing the ore that will achieve the production rate and product quantities and qualities. The processing method must consider the range of physical (such as hardness, abrasiveness or fines percentages), geological (weathered, fresh or transition) and metallurgical (such as flotation characteristics, leachability and recovery) characteristics of the ore within the deposit. The processing methods selected for assessment must be technically and economically viable at the production rates proposed for the deposit, and need to consider the location, site-specific environmental and community issues, and the cost of plant construction and operation.

In assessing a processing method for the deposit, most conventional processing methods (such as crushing and screening, carbon-in-leach / carbon-in-pulp leaching, flotation, dense medium or magnetic separation, or cyclones/classifiers) have a history of implementation and production results in similar deposits, commodities and locations that can support the selection of methods for consideration. Where the processing method being considered is not conventional and does not have that history of production to rely on, consideration must be given to how sufficient confidence can be gained in that proposed non-conventional processing method to achieve the required level of accuracy and confidence using that method of processing.

Underlying any analysis of processing should be a comprehensive range of metallurgical tests on samples of the deposit that are representative of the material to be processed. For some deposits, this may require bulk samples. This is discussed further in Chapter 5, Metallurgy.

Outputs for both the Pre-Feasibility Study and Feasibility Study should be in such a form that an economic evaluation leading to a cash flow analysis can be calculated.

Ground and surface water management

An understanding of ground and surface water systems is important for a mine plan as they determine:

- practicality of the mining operation if surface and groundwater is not or cannot be managed appropriately
- cost of blasting, as ammonium nitrate fuel oil (ANFO) may need to be replaced by emulsion or similar 'wet' explosives where there are unmanaged water inflows
- drawdown profiles used as inputs to the geotechnical model to determine open pit slope stability
- water inflow to the mine and the volume that needs to be managed by the mining operation
- the number and size of pumps and specification for the pipework required to manage water captured in dewatering bores or sumps (capital and operating cost)
- the type, number, size and specification for surface water control structures (such as diversions, channels or settling ponds) required to manage water run-off from rainfall events
- the contribution to the overall site water balance
- location of accesses and infrastructure
- safety of the operations.

A Pre-Feasibility Study should include development of a conceptual hydrogeological model, informed by available historical records for groundwater monitoring bore level measurements, measured or predicted usage and abstraction rates, and rainfall records, supplemented by pump testing from existing groundwater bores.

A numerical hydrogeological model should be developed to allow preliminary estimation of the mine water balance, and a conceptual plan developed to manage surface water and mine dewatering from both groundwater inflow and surface water run-off. The description, size and location of facilities would be expected and calculation of capital and operating costs. The study needs to include discussion on how surface water run-off and groundwater infiltration to the mine will be managed, and if it has the potential to impact surface water and groundwater flows in the surrounding areas. The need to have pit slope depressurisation strategies in place, if dealing with wet/saturated slopes, should be assessed and discussed.

A Feasibility Study should include a more detailed hydrogeological numerical model, informed by specific pump testing of planned dewatering and water supply bores and the most current geological understanding, to provide a higher level of confidence than the conclusions and predictions of the Pre-Feasibility Study. A groundwater and surface water management plan should be developed, and capital and operating costs estimated at Feasibility Study level of accuracy. Surface water management structures should be informed by mapped catchment data and supported by engineering designs. The updated hydrogeological model should be used to predict how surface water run-off and groundwater infiltration to the mine will interact with surface and groundwater flows in the surrounding areas and be used as input to environmental management. The strategies to achieve pit slope depressurisation, if dealing with wet/saturated slopes, should be identified and included in mining method plans, equipment requirements, and capital and operating cost estimates.

The groundwater and surface water management plan output needs to be in a form suitable for estimating costs to the appropriate level of accuracy, determining whether water use and water capture can be balanced, and appropriate environmental and social outcomes can be achieved.

Ground support and geotechnical stability

A mine plan at any level of study needs to be based on a mine design that is geotechnically robust. To ensure geotechnical stability, a well-developed understanding of the ground conditions is necessary. This understanding is achieved through the development of a geotechnical model.

The geotechnical model is the foundation for any open pit or underground design and is the collective output of four main components: the geology, structure, rock mass and hydrogeology models. The development of a geotechnical model is an evolving process, from Scoping Study to Feasibility Study, as data and analysis gaps and uncertainties are addressed.

The geotechnical model is used to identify ground conditions of similar characteristics (geotechnical domains). These domains are based on the rock mass strength, geological structure and potential failure modes on which the mine design is developed.

The geotechnical design is also influenced by external factors such as mine regulations, appropriate mine design configurations, dependent on production requirements and equipment planned for the operation. Based on these requirements, various mine design configurations are tested in geotechnical stability assessments.

Slope stability analysis for open pits involves batter, inter-ramp and overall slope stability assessments. The main inputs are the rock structures (such as faulting and/or joints), rock mass strength and groundwater influence, although groundwater influence is a variable that can potentially be controlled by depressurisation and dewatering. Structural controls may also be mitigated to some extent. Kinematic slope stability analysis provides a basis for optimising bench and inter-ramp scale design (bench face angle, bench height and safety berm width), by assessing the potential failure imposed by the interaction of defects sets for a given wall orientation. The orientation of the defects, their continuity and persistence, and their surface characteristics (shear strength) are the main considerations. The structural information can be obtained from outcrop mapping in the study phase. The structural logging of oriented drill holes will provide limited information, such as the orientation of the defects, but no information on the continuity or persistence. Therefore, assumptions are made for these parameters in the designs. Those assumptions have to be verified in the operational phase.

Guidelines showing five levels of geotechnical model development, depending on various phases of data collection and data confidence, were developed by Read and Stacey (2009). They are qualitative in nature, although numerical confidence levels have been defined (see Table 4.2).

The level of confidence of geotechnical modelling relative to the JORC Code is shown in Figure 4.2.

Mining method selection

Once the scale of mining or the range of production rates to be assessed has been defined, consideration must be given to the methods of mining that will achieve the production rate or rates identified. The mining method must consider the characteristics of the deposit, be technically and economically viable at the production rates proposed for the deposit and needs to consider the scale and geometry of the mineralisation, location, site access, environmental and community issues, and the capital cost of implementation. If equipment automation is a consideration, that needs to be factored in at an early stage.

TABLE 4.2

Suggested target levels of data confidence by study phase (Dunn, 2014, after Read and Stacey, 2009*).

PROJECT PHASE					
Project level status	**Conceptual**	**Pre-Feasibility**	**Feasibility**	**Design and construction**	**Operations**
Geotechnical level status	**Level 1**	**Level 2**	**Level 3**	**Level 4**	**Level 5**
Geotechnical characterisation	Pertinent reginal information	Assessment and compilation of initial mine scale geotechnical data	Ongoing assessment and compilation of all new mine scale geotechnical data	Refinement of geotechnical database and 3D model	Ongoing maintenance of geotechnical database and 3D model
Target levels of data confidence for each model					
Geology	>50%	50–70%	65–85%	80–90%	>90%
Structural	>20%	40–50%	45–70%	60–75%	>75%
Hydrogeological	>20%	30–50%	40–65%	60–75%	>75%
Rock mass	>30%	40–65%	60–75%	70–80%	>80%
Geotechnical	>30%	40–60%	50–75%	65–85%	>80%

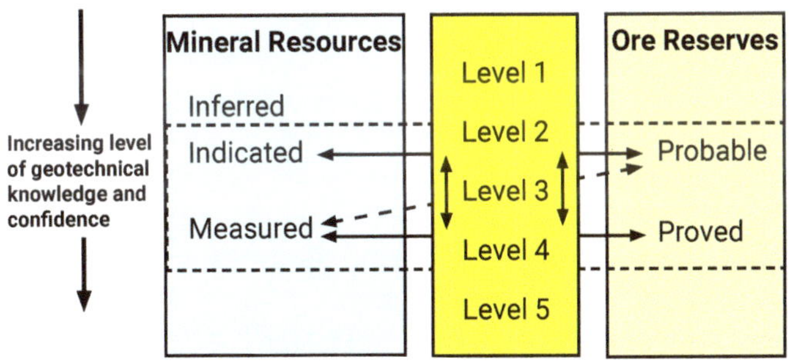

FIG 4.2 – Geotechnical levels of confidence relative to the JORC Code (after Read and Stacey, 2009*).

* Reproduced from *Guidelines for Open Pit Slope Design*, Read and Stacey (2009) with permission from CSIRO Publishing.

In assessing a mining method for the deposit, most conventional mining methods (such as underground, open pit truck and shovel, dragline, or dredge mining) have a history of implementation and production results in similar deposits, commodities and locations. Where the mining method or method of extraction being considered is not conventional and does not have that history of production to rely on, consideration must be given to how sufficient confidence can be gained in that proposed non-conventional mining method to justify the appropriate accuracy required of the level of study (and for a Reserve estimate) using that method of extraction.

There are too many mining methods to discuss, but the essential themes of the available methods are to decide on whether the mining method is, for example:

- surface, underground, dredging, *in situ* or undersea
- conventional mining (history of performance) or innovative techniques (little or no history)
- new technology (automation, optical sorters) or traditional (excavator and haul truck)
- selective mining (small equipment) or bulk mining (larger equipment)
- draglines, dozer trap or truck and shovel (coal)
- naturally supported, artificially supported, caving or *in situ* (underground)
- drill and blast, free dig or mechanical breakage (surface miner or bucket wheel excavator).

A Scoping Study might assess a broad range of mining methods at a high level to identify the practical and feasible methods to be studied in more detail in the Pre-Feasibility Study. The focus at Scoping Study phase is to identify a feasible method rather than the optimum method.

A Pre-Feasibility Study might assess a number of mining methods to identify the preferred method. The Pre-Feasibility Study should provide a description of the mining methods considered, conditions under which the proposed mining method is technically viable, the mining equipment required to undertake the mining method as well as the support and ancillary equipment, personnel, and infrastructure required to support the mining method. It will need to discuss, per JORC Code Table 1, Section 4 Mining Factors or Assumptions:

- the method and assumptions used as reported in the Pre-Feasibility or Feasibility Study to convert the Mineral Resource to an Ore Reserve (ie either by application of appropriate factors by optimisation or by preliminary or detailed design)
- the choice, nature and appropriateness of the selected mining method(s) and other mining parameters including associated design issues such as pre-strip, access, etc
- the assumptions made regarding geotechnical parameters (eg pit slopes, stope sizes, etc), grade control and pre-production drilling
- the major assumptions made and Mineral Resource model used for pit and stope optimisation (if appropriate)
- the mining dilution factors used
- the mining recovery factors used
- any minimum mining widths used
- the manner in which Inferred Mineral Resources are utilised in mining studies and the sensitivity of the outcome to their inclusion
- the infrastructure requirements of the selected mining method.

A Feasibility Study is likely to assess only the preferred mining method identified from the Pre-Feasibility Study in more detail. The Feasibility Study should provide a more detailed description of the mining method, its application in the specific conditions of the deposit, the size and number of mining equipment required to undertake the mining method as well as the support and ancillary equipment, personnel and infrastructure required to support the mining method.

Outputs for both the Pre-Feasibility Study and Feasibility Study should be in such a form that a mining fleet evaluation leading to a mining cost estimate can be applied.

Mine design and infrastructure layout

The mining inventory used for mine scheduling and evaluation in a study (and a Reserve) needs to be based on a mine design that has been developed with regard to the specific characteristics of the deposit being evaluated. As a guide, the JORC Code states in the guidance notes that 'Deriving an Ore Reserve without a mine design or mine plan through a process of factoring of the Mineral Resource is unacceptable'.

The mine design should present a 3D spatially correct design in electronic format showing the economic limits of extraction, means of access for ore, waste, personnel, materials and equipment, and be in a form that can be set out in the field by a qualified surveyor using the spatial information in the mine design. The mine design should form the basis for estimating quantities and qualities of ore, and waste material.

The mine design is informed by a wide range of considerations that will be different depending on the mining method, commodity and location. Common to all will be the resource model, geotechnical design criteria, access considerations, and other influencing factors such as hydrogeology, economic inputs and mining selectivity. Automation of mining equipment may require a different mine design to manual operations.

The mine design must account for ESG issues, such as restrictions placed on mine development by, for example:

- threatened habitat, sensitive environmental areas, drainage or water supply areas
- water abstraction and disposal
- areas of cultural, religious or heritage significance
- local communities and community infrastructure, such as residences, bores, roads and utilities.

In addition, the mine design must account for practical and safety considerations to safeguard key infrastructure at the mine, such as the process plant, tailings storage facility, workshops, power lines, access roads and services corridors, although infrastructure trade-off studies may determine that it is more economic to shift or rebuild infrastructure than to lose access to ore that might be sterilised by the presence of the infrastructure.

A Scoping Study should identify the key mine design and infrastructure elements and address, at a conceptual level, the requirements of those key elements, which will be addressed in more detail in the Pre-Feasibility Study. A mine layout should be presented, with the focus on identifying a feasible mine layout rather than undertaking the detail and trade-off studies required to define the optimum mine layout.

A Pre-Feasibility Study mine design has, according to the JORC Code, '… advanced to a stage where a preferred mining method, in the case of underground mining, or the pit configuration, in the case of an open pit, is established …'.

Thus, for an underground mine:

- the mining method needs to be identified
- stope (and if required, pillar) configurations should be established for that mining method

- backfilling methodology (if applicable), mine development design and profiles should be prepared
- means of access and egress for ore, materials and personnel, key infrastructure for ventilation and other mine services, materials handling, and personnel safety should be established.

And for an open pit mine, pit optimisation shells or floor contours would be insufficient to define the pit configuration, which requires the details of haulage ramp alignments from which haulage ramp gradient and width can be established, assessment of minimum working widths, and batter and berm configurations from which batter heights, slope angles and berm widths can be established.

A Feasibility Study mine design should progress the level of detail and accuracy of a Pre-Feasibility Study mine design so that the level of confidence in the design is increased to the point that cost estimation is within the required order of accuracy.

Option analysis

At Scoping Study level, generally only one option is assessed, and as long as that option demonstrates the potential for technical and economic viability, progression to a Pre-Feasibility Study is justified, at which time alternative options can be assessed.

Option analysis is undertaken in the Pre-Feasibility Study to assess the options for developing the project to identify the preferred project development pathway. Options available will depend on the scale of the mineralisation, orebody characteristics, location, environmental and community issues, marketing analysis, and the cost of implementation.

At Feasibility Study level, the preferred development options are assessed in more detail with additional engineering rigour applied.

All of the areas under the headings of this chapter can be included in options analysis, with some of the more common areas being processing and mine development strategy. Processing options are discussed under processing method selection.

Mine development strategy options are generally undertaken using specialist optimisation software, which may include pit optimisation (open pit), mineable shape optimisation (underground) or mining schedule optimisation software for analysis of mine sequencing or cut-backs (open pit). Strategy optimisation may involve assessment of different options for such issues as:

- *production rates* – see discussion under the 'Mine scale' section in this chapter
- *cut-off grades or values* – in combination with production rates
- *mining methods* – see discussion under 'Mining method selection'
- *mining equipment* – alternative fleet configurations, automation or manual operations, electric, battery or diesel equipment
- *access requirements for underground mines* – declines, adits or shafts/internal shafts
- *deposit development strategies* – cut-back sequence or dredge path in an open pit mine, longwall or stope extraction sequence in an underground mine
- *material handling options* – in-pit or underground crushing and conveying, shaft hoisting or haul trucks.

Option analysis needs to be done at a sufficient level of detail and engineering analysis to provide confidence that the right options are selected and that an appropriate Pre-Feasibility Study level of assessment was achieved to support a decision of the preferred option.

Mining equipment selection

At Scoping Study level, mining equipment selection is typically not addressed, except in a general sense by applying assumptions to the mine plan for selectivity, equipment productivity and operating costs that are applicable to a specific size and type of equipment.

For a Pre-Feasibility or Feasibility Study, consideration of the appropriate mining equipment is required. Mining equipment selection is an iterative process that follows on from selection of the scale of mining, the mining method and the mine design. It is likely that some level of trade-off study has been undertaken to arrive at the selected fleet configuration, as the range of options is very diverse and both practicality and economic assessment is involved in the selection.

An equipment fleet may be assumed to develop mining cost inputs to the initial mine planning phase of a study. A second iteration of equipment fleet production capacity may be required once detailed mine design has identified mining quantities higher or lower than initially assumed from the results of previous less-accurate studies. An additional iteration of equipment fleet may be required after mine scheduling identifies that higher or lower equipment capacity is required to achieve scheduling targets.

Equipment fleet analysis requires assessment and selection (size rather than specifying make and model) for each component in the mining cycle:

- *automation versus manual operation* – this needs to be discussed early in the process as automation may require a different mine plan and mine layout
- *cutting* – continuous miners, road headers, tunnel boring machines, surface miners, longwall shearers, dredges or hydraulic miner
- *loading* – draglines, rope shovels, hydraulic excavators, front-end loaders
- *haulage* – haul trucks (such as on-road, off-road rear-dump, off-road side-tippers, all terrain or specialised haulage units), in-pit crushing and conveying, conveyors (such as conventional, high lift, flyte or rope conveyors), slurry pipelines, rail or underground mining hoisting shafts
- *drilling* – surface rotary, top-hole or down-the-hole-hammer drills, development jumbos, underground longhole production drills, raise borers or other specialised drills
- *ancillary equipment* – tracked and rubber-tyred dozers, motor graders, water trucks, front-end loaders, service trucks, fuel trucks, charge-up units, mobile manufacturing units, concrete sprayers, secondary breaking units and agitators
- *support equipment* – integrated tool carriers, bobcats, forklifts, tyre changers, lighting plants, pumps, buses, light vehicles, cranes.

All of this equipment has a long history within the mining industry and production performance, maintenance requirements, operating costs and purchase costs are well known and widely available, with most mining companies and consulting firms having access to a database of productivity and cost estimates for a wide range of equipment from a range of studies in different commodities and locations, as well as from original equipment manufacturers' estimates.

However, continuing developments in battery technology may also bring conventional mining equipment powered by batteries into consideration, which do not have that same long history of performance. Consideration needs to be given as to how the appropriate study level of accuracy can be achieved. Similarly with automation of mining fleets or any other innovative mining equipment.

A Pre-Feasibility Study might be considering multiple equipment fleet configurations to identify the preferred trade-off between production rates and overall operating (including allowance for ownership) costs. The Pre-Feasibility Study should identify the prime production tools that determine mine production rates such as cutting tools, loading tools, haulage units and drilling units.

Size of mining equipment and productivity and operating costs estimates derived from similar projects or a database of equipment data should be specified. In owner mining options, ownership costs should be estimated. Ancillary and support equipment should be listed, but detailed analysis is not warranted to achieve an appropriate Pre-Feasibility Study level of accuracy.

A Feasibility Study looks in more detail at a single equipment fleet configuration, although there may be multiple equipment sizes within that fleet for different production roles, such as a smaller digger and trucks for ore than there may be for waste, which might be mined in larger quantities on larger benches. The Feasibility Study should identify the prime production tools as well as the ancillary and support equipment requirements. Capital and operating costs should be supported by manufacturer or distributor quotes to support an equipment cost estimate at Feasibility Study level of accuracy.

Alternatively, mining fleet requirements and mining cost estimates could be provided by a mining contractor budget estimate (Pre-Feasibility Study) or a quote capable of commercial acceptance (Feasibility Study) based on the mine design and mining schedule provided by the company.

Mining infrastructure and services

Appropriate infrastructure to support the operation and the mine plan needs to be available either locally or provided by the mining company or its service providers (such as mining contractors). Often called non-process infrastructure, a description and specification of the infrastructure required to support the operation is required to estimate the capital cost of construction, sustaining capital cost of maintenance, and operating cost of the infrastructure and services (power, water, gas and telecommunications) required to mine and transport products to market.

The infrastructure required will depend on the scale of operations, commodity, mining and processing methods used, location, and the existence of appropriate locally available infrastructure. For example, a fly-in, fly-out operation requires access to an airstrip and an accommodation village that would not be required if there were a local workforce. Infrastructure such as crushers and conveyors may be required where material handling issues, particularly for large volume bulk commodities such as coal, bauxite or iron ore, make it difficult or more expensive for haul trucks to transport ore to a centrally located processing plant and products to a port or warehouse facility.

The availability of land for infrastructure development and approvals from regulatory authorities may determine the scale of cost of providing infrastructure. Power, water and telecommunications are often available locally, so that connection to the local grid is all that is required. However, often power and water needs to be provided for remote minerals projects. And while a power station can always be constructed as part of a project to deliver power, water availability of suitable quality can be a critical component, without which the project cannot proceed.

A Scoping Study may not necessarily consider these issues in detail, however, appropriate allowances for capital and operating costs should be made and a risk assessment undertaken to identify issues.

A Pre-Feasibility Study should provide a sufficient description of the size and type of infrastructure required so that an estimate of capital cost and sustaining capital cost can be provided. Such estimates often use contingency estimates to allow for the unknown additional costs and lack of definition. Costs should be developed from supplier budget estimates or estimates of similar infrastructure provided for other similar projects and should achieve the appropriate Pre-Feasibility Study level of accuracy.

A Feasibility Study should provide a detailed description and engineering specification and drawing of the infrastructure required so that a first principles estimate of capital cost can be developed from material offtakes and labour costs or a quote obtained from a provider. Sustaining capital cost should be estimated from first principles and cross-checked with experience from other similar infrastructure at other projects. Contingency provisions are smaller as higher confidence cost estimates are developed.

Pre-Feasibility and Feasibility Study estimates should be in a form suitable for estimating capital and operating costs and allow discussions on approvals from regulatory authorities and local communities affected by the development.

Waste management

There should be consideration of a plan to manage the waste material generated by the mine at the earliest stage, including the requirements for mine closure. For an open pit mine, waste is stored in an ex-pit or in-pit waste landform or ex-pit and in-pit spoil piles (draglines). For an underground mine, waste generation is less of an issue as waste rock is only generated from development and may be used within the mining operation as backfill, although some small waste dumps are generated from initial mine development and excess development waste.

In-pit waste landforms are often preferred where they don't sterilise orebody extensions or ore at depth. Waste landforms, whether in-pit or ex-pit, need to be designed specifically to the material characteristics of the waste mined and the mine closure requirements. Design parameters include geotechnical parameters (overall slope, bench height, batter angle and berm width) to design sustainable and stable batter slopes that can support revegetation, with berms to allow for drainage and slope breaks, landform shapes that fit in with the local topography, and a conceptual cover sequence design for the top of the dump to either shed or capture rainfall, depending on the requirements of the local environment and vegetation, permitting requirements and the mine closure design.

In addition, mines may produce waste products that require special deposition or encapsulation. For example, sulfide, radioactive, fibrous or combustible materials. Open pits that mine fresh rock often contain waste rock with sulfide mineralisation that has the potential to generate acid and may leach metals into the environment.

At a Scoping Study level, a strategy and cost allowance for waste management is required, and the issues discussed for a Pre-Feasibility Study may not be known.

A Pre-Feasibility Study should address acid mine drainage (AMD) issues and undertake a waste rock characterisation study to identify the rock types that have AMD potential and the design parameters required to manage any adverse effects. Potentially acid-forming (PAF) waste rock will need to be encapsulated within a non-acid forming (NAF) enclosure to prevent acid or metal leachate escaping into the groundwater, local streams or the environment. A specific geochemical waste sampling program should be undertaken and static test work undertaken, with kinetic test work in progress. The likely mine closure requirements should be identified and addressed in the mine plan and a conceptual mine closure plan undertaken.

A Feasibility Study should have waste dump designs informed by the results of the geochemical kinetic test work and the preliminary mine closure plan, and additional work undertaken to:

- identify and plan removal of unsuitable waste materials at the base of the waste landform
- undertake net carbonate value modelling within the resource model
- schedule PAF and NAF waste type presentation in mine scheduling to ensure there is sufficient NAF material to manage PAF material
- design the waste dump to contain PAF waste within a layer of NAF material
- generate the waste landform designs to stable sustainable overall slopes suitable for revegetation and in keeping with topographic and mine closure parameters.

The waste plan should be in a form suitable for cost estimation to the appropriate study level of accuracy and needs to include a conceptual closure plan for the landform and proposed end use for the land.

Tailings management

A full discussion of the requirements for tailings management is provided in Chapter 7.

Mining and processing schedule

The mining and processing schedule is the key to taking the material quantities and qualities in the mine design and providing a time-based estimate of when revenue will be available, and costs incurred over the life of the mineral asset to generate discounted cash flows. The schedule provides the physical estimates in ore and waste tonnes mined and processed, stockpile movements, feed qualities, and product and contaminant quantities and qualities. This allows a work-up estimate of revenues and costs by time period (see mining revenue and cost estimation).

Mining schedules are often iterative, with a first pass schedule determining approximate quantities so that mining equipment fleet analysis can be undertaken and an equipment fleet determined. Once the equipment fleet is determined, the schedule can be re-run to optimise the utilisation of the mining fleet identified from the first pass mining schedule. This type of resources-based scheduling allows mining fleets to be operated more efficiently and mining costs minimised.

Different commodities and different mining methods are scheduled in different ways, depending on the characteristics of the deposit and the mining method. More systematic mining methods where it is difficult to relocate mining equipment and its associated infrastructure (such as in-pit crushing and conveying, dozer traps, large bucket wheel excavators or draglines) are less flexible than a highly mobile equipment fleet such as front-end loaders and haul trucks.

A number of strategic scheduling software packages are available and widely used to assist in optimising the strategy behind a mining schedule, and can flex key drivers such as production rates, cut-off values, product qualities, start and finish dates by mining areas, stockpiling strategies, and mining advance rates to deliver improved economics.

A Scoping Study mining and processing schedule would generally be at a high level and in longer time periods than required for a Pre-Feasibility or Feasibility Study and may provide the strategy for the Pre-Feasibility Study scheduling.

A Pre-Feasibility Study should develop a detailed mining area by mining area schedule to identify ore and waste tonnages by period, as well as ore qualities, processing feed tonnes and qualities, and stockpile movements. For underground mines, this would also include key physical parameters such as lateral and vertical development metres, production drilling metres, backfill volumes and haulage requirements. Time periods should be shorter towards the start of the schedule (possibly quarterly for two to five years) as more detailed information is required and longer towards the end of the schedule (possibly annual), depending on the length of the schedule. For long-life operations, +30 years, five-year time periods towards the end of the schedule might be appropriate.

A Feasibility Study should develop a detailed mining area schedule based on equipment numbers and productivities to identify ore and waste tonnages by period, as well as ore qualities, processing feed tonnes and qualities, and stockpile movements. For underground mines, this would also include key physical parameters such as lateral and vertical development metres, production drilling metres, backfill volumes and haulage requirements. Time periods should be shorter towards the start of the schedule (usually monthly for two to five years) as more detailed information is required and longer towards the end of the schedule (possibly annual). This information would be used to estimate equipment hours and construct a mining cost model with haulage data.

Mining revenue and cost estimation

To determine that a Reserve is economic, both the revenue generated from that Reserve and the cost of generating that revenue (see Chapter 9) up to the point of sale should be estimated. This will require the following components to have been completed:

- the mine design to generate physical quantities of ore and waste and qualities of the ore
- the mining and processing schedule to determine the time-based distribution of ore and waste movements and ore qualities
- processing assessment to determine the processing method, throughput, metallurgical recovery performance, product and contaminant qualities
- capital cost of construction and sustaining capital cost of maintaining infrastructure
- operating cost, including all site (mining, processing and administration) costs and applicable off-site costs
- mine closure costs
- the key economic and cost parameters collated in the basis of design, discussed under mine planning criteria.

If Inferred Resources are utilised in the mine plan, the JORC Code states that:

> *Confidence in the estimate of Inferred Mineral Resources is not sufficient to allow the results of the application of technical and economic parameters to be used for detailed planning in Pre-Feasibility (Clause 39) or Feasibility (Clause 40) Studies.*

Therefore for external reporting, Inferred Resources, if included in the mining and processing schedule, need to be treated as having no value in the economic assessment used to justify extraction of the Reserve. The Reserve alone should support a project with positive economics.

A Scoping Study may include Inferred Resources in the mine plan and economic assessment, as a Reserve cannot be publicly reported and is generally not estimated. The economic outcome generally provides justification to proceed to a Pre-Feasibility Study.

A Pre-Feasibility Study should discuss (JORC Code Table 1, Section 4 under Costs, Revenue Factors and Economic):

- the derivation of, or assumptions made, regarding projected capital costs in the study
- the methodology used to estimate operating costs
- allowances made for the content of deleterious elements
- the source of exchange rates used in the study
- derivation of transportation charges
- the basis for forecasting or source of treatment and refining charges, penalties for failure to meet specification, etc
- the allowances made for royalties payable, both to government and private entities
- the derivation of, or assumptions made, regarding revenue factors including head grade, metal or commodity price(s) exchange rates, transportation and treatment charges, penalties, net smelter returns, etc
- the derivation of, or assumptions made, regarding metal or commodity price(s), for the principal metals, minerals and co-products
- inputs to the economic analysis to produce the net present value (NPV) in the study, the source and confidence of these economic inputs including estimated inflation, discount rate, etc
- NPV ranges and sensitivity to variations in the significant assumptions and inputs.

A Feasibility Study needs to address the same aspects, but at a higher level of accuracy.

Mine closure planning

Mine closure to achieve a stable environmental outcome agreeable to all stakeholders in a project is a significant component of the mine plan and social licence to operate. The timing for commencement of closure works and the cost estimate may have a significant impact on the economics of a short-life project.

Preparation of mine closure plans are typically incorporated and undertaken as part of the environmental permitting components discussed in Chapter 2.

Increasingly, mineral asset approvals and financing require more detailed ESG assessment and mine closure plans must be considered in the Modifying Factors for a Reserve. Closure planning will be a work in progress throughout the life of the mineral asset and different trials for stable landform design and management, landform cover design, vegetation design, seed mix, habitat construction and contouring profile will inform the final mine closure plan much closer to the end of mine life. However, a conceptual closure plan is required prior to commencement, so that mine closure principles are embedded in mine design from the start. This is likely to include tailings management landforms and waste landform locations, heights and profiles, drainage and diversion designs, and topsoil management. In jurisdictions where a rehabilitation fee/levy and/or bond is required, this amount should be included in economic models.

A Scoping Study is likely to include a cost allowance derived from other similar projects. A Pre-Feasibility Study would need to consider a conceptual mine closure plan and cost estimate for returning the mine and associated infrastructure areas to the agreed standard on mine closure.

Issues considered would be waste rock characterisation and AMD issues, allowance for ongoing rehabilitation, and allowing the time and cost for mine closure in the schedule.

A Feasibility Study would need to consider a more detailed, but still conceptual, mine closure plan underpinned by a spatially correct and engineered landform and drainage layout design. This would allow material quantities and costs to be more accurately estimated to rehabilitate the mine and associated infrastructure areas.

Mine closure plans should be in a form suitable for estimating a cost at the required level of accuracy for the study phase and determining whether appropriate environmental and social outcomes can be achieved.

Estimating Reserves

A Scoping Study is not considered under the JORC Code to be a reasonable basis for an Ore Reserve, which requires a minimum of a Pre-Feasibility Study level of assessment. The level of confidence is aimed at what the operation or proposed operation being assessed can achieve, so there may be a different focus for achieving that level of confidence for an operating mine (life-of-mine plan and budget) than for a greenfield project (Pre-Feasibility Study or Feasibility Study report), but the requirement for a minimum of a Pre-Feasibility Study level of assessment of Modifying Factors does not change.

Although the level of a study dictates the detail in which mining aspects are assessed, when classifying an Ore Reserve, the levels of confidence of the underlying Mineral Resource and the Modifying Factors (as defined in the JORC Code) determine the classification level of the Ore Reserve, not the level of the study. Whether a Pre-Feasibility Study or Feasibility Study is being undertaken, the level of assessment of Modifying Factors needs to meet the level of confidence required by each classification level. Both Proved and Probable Ore Reserves can be reported from a Pre-Feasibility Study or a Feasibility Study.

However, it is expected that the confidence in the Modifying Factors for a Feasibility Study would, in general, be greater than the confidence level for a Pre-Feasibility Study. If done correctly, the amount of Reserve may not change between a Pre-Feasibility Study and a Feasibility Study, but the level of confidence in the Modifying Factors may increase and that may or may not impact the classification level of parts of the Reserve and the relative project value between the two phases of study.

The JORC Code

An Ore Reserve should not be confused with a mineral inventory. Mineral inventory is not a defined term in the JORC Code but can be used to refer to all mineralisation regardless of Resource classification that has been included in the study design and schedule. Only the Measured and Indicated components of the Resource can be converted to a Reserve through the application of Modifying Factors determined during the study process. The JORC Code describes a Reserve (Clause 29), and most other recognised mineral reporting codes provide a similar definition, as:

> *An 'Ore Reserve' is the economically mineable part of a Measured and/or Indicated Mineral Resource. It includes diluting materials and allowances for losses, which may occur when the material is mined or extracted and is defined by studies at Pre-Feasibility or Feasibility level as appropriate that include application of Modifying Factors. Such studies demonstrate that, at the time of reporting, extraction could reasonably be justified.*

The reference point at which Reserves are defined, usually the point where the ore is delivered to the processing plant, must be stated. It is important that, in all situations where the reference point is different, such as for a saleable product, a clarifying statement is included to ensure that the reader is fully informed as to what is being reported.

Applying Modifying Factors to the mine plan

After consideration of all of the above areas, estimation of the Reserve and classification into the appropriate confidence levels in tabular format is required, along with notes critical to understanding the Reserve.

The underlying Resource classification provides the highest Reserve classification that can be assigned, but consideration of all of the Modifying Factors is required to determine the level of confidence of the Reserve. If all material Modifying Factors are known to the same level of confidence as the underlying Resource, Measured Mineral Resources can be converted to Proved Ore Reserve and Indicated Mineral Resources to Probable Ore Reserve (see the horizontal arrows in Figure 4.1). However, if any of the material Modifying Factors are not known to the same level of confidence as the Resource, the Measured Mineral Resource may be downgraded to Probable Ore Reserve and the diagonal arrows in Figure 4.1 apply.

There are decisions made during the study that could affect the outcome of the Reserve, so there is no unique 'correct' Reserve. A company's strategy may make a larger Reserve with lower economic value more attractive than a smaller Reserve with higher economic value.

In addition, there are a number of non-technical Modifying Factors to be considered that can affect whether a Reserve can be estimated and the quantity and classification of the Reserve, as follows.

Permitting

The status of permitting to operate the mining and processing operation is a Modifying Factor and needs to be considered. Regardless of the level of technical and economic assessment undertaken, if permission to mine a mineral asset is unlikely to be granted, a Reserve cannot be estimated. The impact of the following on the estimated Reserve for a mineral asset must also be considered and adjustments made as required:

- Any identified material naturally occurring risks.
- The status of material legal agreements and marketing arrangements.
- The status of governmental agreements and approvals critical to the viability of the project, such as mineral tenement status, and government and statutory approvals. There must be reasonable grounds to expect that all necessary government approvals will be received within the time frames anticipated in the Pre-Feasibility or Feasibility Study. Highlight and discuss the Materiality of any unresolved matter that is dependent on a third party on which extraction of the Reserve is contingent.

Environmental factors

The status of studies of potential environmental impacts of the mining and processing operation are a Modifying Factor and need to be considered, along with details of waste rock characterisation studies and the consideration of potential sites, status of design options considered and, where applicable, the status of approvals for tailings storage and waste dumps. In many cases, environmental factors are linked to permitting factors and this needs to be considered when determining how likely it is that approvals will be received in the time frame envisioned by the mine plan.

Social factors

The status of agreements with key stakeholders and matters leading to social licence to operate the mining and processing operation are a Modifying Factor and need to be considered. In some cases, social factors are linked to permitting factors and this also needs to be considered when determining how likely it is that approvals will be received in the time frame envisioned by the mine plan.

Market factors

Marketing analysis will need to assess:

- the demand, supply and stock situation for the particular commodity, consumption trends and factors likely to affect supply and demand into the future
- a customer and competitor analysis along with the identification of likely market windows for the product
- price and volume forecasts and the basis for these forecasts
- for industrial minerals, the customer specification, testing and acceptance requirements prior to a supply contract.

Competent Persons

Although an Ore Reserve can be estimated by a suitably qualified mineral industry professional, Public Reporting requires that (per the JORC Code):

> *Documentation detailing Exploration Results, Mineral Resource and Ore Reserve estimates, on which a Public Report on Exploration Results, Mineral Resources and Ore Reserves is based, must be prepared by, or under the direction of, and signed by, a Competent Person.*

The definition of a Competent Person is included in the JORC Code and may be updated from time to time. This must be considered when assembling a study team to ensure that a suitably qualified and experienced person or persons (there may be more than one Competent Person) are available in the study team. Where there are clear division of responsibilities in a study, it is usually preferable to have multiple Competent Persons sign off their individual areas of responsibility.

For example, deposits with complex metallurgy may require a suitably qualified and experienced metallurgist to be Competent Person for the ore processing aspects of a Reserve. Similarly for other specialised functions in complex mineral projects where particular expertise is required, such as tailings management, geotechnical assessment, marketing or financial evaluation. Each Competent Person must be realistic about their qualifications in accepting Competent Person responsibility and need to satisfy the JORC Code definition of a Competent Person.

The requirement for an Ore Reserve under the JORC Code that 'Such studies demonstrate that, at the time of reporting, extraction could reasonably be justified' provides significant challenges for emerging industries and rapidly changing technologies that seek to exploit mineral assets that have different characteristics and may lie in different environments to more traditional mineral assets. In the absence of a history of performance and costs at similar projects, how does the Competent Person demonstrate that they have made 'reasonable assumptions' on the Modifying Factors? In these cases, it may not be possible to find suitably qualified and experienced personnel who meet the definition of a Competent Person and multiple Competent Persons are likely to be required.

Examples of this may be extraction of minerals from evaporites or brines, recovery of deep-ocean sea floor metallic nodules, or mining of previous niche minerals that may have widespread adoption in emerging technologies, such as was the case for lithium, graphite and rare earth minerals.

CONCLUSIONS

The mining aspects of mineral projects are assessed in increasing levels of detail as studies progress from Scoping Study through to Feasibility Study. Each study phase needs to consider multiple aspects to achieve the required level of accuracy for that study. However, generally speaking, the earlier each aspect is considered in a study, the more appropriate the mine design will be to account for that aspect. As an example, the earlier that issues such as automation, mine closure and ESG requirements are considered in the mine plan, the more likely it is that the mine plan will make the appropriate allowances and allow mineral company management to be informed of specific requirements.

Each study must establish mine planning criteria and undertake the mine planning and assessment work to evaluate all of the issues affecting the mine plan to the required level of confidence for that phase of study, which includes both technical and non-technical issues.

For an Ore Reserve to be estimated and reported from the outcome of a study, the study must be at Pre-Feasibility or Feasibility Study level and the Modifying Factors must be assessed to the required level of confidence appropriate to the classification of Reserve. The study team needs to include suitably qualified and experienced Competent Persons, as defined by the JORC Code, and there may be a need for multiple Competent Persons or specialists to provide input at any level of study and into mining operations.

An economic assessment is required to confirm that the mine plan is technically and economically achievable and extraction of the Reserve is justified at the time of reporting using reasonable financial assumptions. Inferred Resources must be considered as having no value in an economic assessment.

ACKNOWLEDGEMENTS

The author wishes to acknowledge the contributions of the following people, including an anonymous peer reviewer:

- J Busfield, Principal Mining Engineer, MineCraft Consulting
- R Chesher, Principal Metallurgical Consultant, AMC Consultants Pty Ltd
- L Gillett, Practice Leader – Corporate/Principal Mining Engineer, AMC Consultants Pty Ltd
- A Herath, Principal Geotechnical Engineer, AMC Consultants Pty Ltd
- C Kraut, Principal Hydrogeologist, Advisian Group
- D Lee, Underground Manager Perth/Principal Mining Engineer, AMC Consultants Pty Ltd
- M Liu, Senior Mining Engineer, AMC Consultants Pty Ltd
- S Kusabs, Principal Mining Engineer, AMC Consultants Pty Ltd
- D Varcoe, Principal Mining Engineer/Open Pit Discipline Lead, AMC Consultants Pty Ltd
- M Walske, Manager Geotechnical Perth/Principal Geotechnical Engineer, AMC Consultants Pty Ltd

REFERENCES

Dunn, M, 2014. Geotechnical models and data confidence in mining geotechnical design, in *Proceedings AusRock 2014: Third Australasian Ground Control in Mining Conference,* pp 127–133 (The Australasian Institute of Mining and Metallurgy: Melbourne).

JORC, 2012. Australasian Code for Reporting of Exploration Results, Mineral Resources and Ore Reserves (The JORC Code) [online]. Available from: <http://www.jorc.org> (The Joint Ore Reserves Committee of the Australasian Institute of Mining and Metallurgy, Australian Institute of Geoscientists and Minerals Council of Australia).

Read, J R L and Stacey, P F, 2009. *Guidelines for Open Pit Slope Design* (CSIRO Publishing: Melbourne).

5

METALLURGY

INTRODUCTION

The metallurgy section of a study continues the development of the fundamental building blocks, which commenced with geology. Metallurgical test work informs mine planning, process options, capital and operating costs, and project feasibility.

In embarking on the metallurgical journey, an understanding of the geology is a necessity, as it provides the basis for metallurgical sample section and preliminary domain identification for what may become, or already be defined, as a Resource.

As metallurgical understanding progresses, domaining from a metallurgical perspective may change and different domains may be defined for different processes, for example comminution, flotation and extractive leach processes.

This chapter is intended to provide guidance on what is addressed in metallurgy (Table 5.1), as well as a high-level list of the types of metallurgical investigations that may be undertaken at each study phase (Table 5.2).

Table 5.2 provides an indication of the detailed test work that may be undertaken and in which study phase.

The test work only applies depending upon the processes being considered. For example, in a whole ore, leach mineral association may not be useful, however mineral liberation may be of use if there is encapsulation occurring that affects potential extraction extents.

Additional discussion in support of the summary tables is found within the chapter. There are many resources available to explain how to undertake test work, but not so much on how to align test work with study phases. Resources include internal company procedures, technical papers, books and online resources.

This chapter is meant to start you in your journey towards understanding what is required in a study phase. Not all processes may be addressed. It is intended that there should be sufficient information in this chapter to enable your independent interpretation of the requirements by study phase for metallurgical processes.

This chapter is intended as a complement to internal and external study reports and is not intended to replace external reporting guidelines, where available and applicable. As will be noted through this handbook, external reports are generally a summary of the work undertaken and reported in an internal study report.

TABLE 5.1

Metallurgy study phase guide. This table is not meant to be an exhaustive list of process areas for investigation, it is merely intended to provide a guide.

METALLURGY					
Activity	**Deliverable**	**Description**	**SS**	**PFS**	**FS**
Sample provenance	Report, including maps	Increasing number and representation of the Resource elements. Resource elements may include, for example: • lithology • alteration • grade (value and deleterious elements) • mineralogy • spatial orientation.	✓	✓	✓
Test work planning	Test work plan	Describe sample selection; test work objectives; proposed test work.	✓	✓	✓
		Flow sheet options.	✓	✓	
Process test work	See Table 5.2	Test work report for process area(s). Ensure increasing confidence in test work parameters, which inform facility design.	✓	✓	✓
Reporting	Test work reports	Test work reports for process area(s), documenting scope of work, sample provenance from test work facility perspective, test work procedures, test work results and interpretation where applicable. Key primary data documents as basis for interpretation.	✓	✓	✓
Test work interpretation	Process design criteria (derived from test work reports)	Parameters for equipment sizing, which may be assumed in the Scoping Study.	✓	✓	✓
		Recovery definition.	✓	✓	✓
		Geometallurgy.			✓
		Metallurgical domaining.	✓	✓	✓
Flow sheet development	Option assessment(s)	Base documents include test work reports, and test work interpretation. Consider potential flow sheet options for Resource. Continue into Pre-Feasibility Study only if outstanding/incomplete from Scoping Study. May be specific to capacity or equipment selection, for example in Pre-Feasibility Study.	✓	✓	
	Process flow sheet		✓	✓	✓
Gap analysis	Future workplan	Identifying areas for additional investigation.	✓	✓	✓

SS = Scoping Study; PFS = Pre-Feasibility Study; FS = Feasibility Study.

TABLE 5.2

Metallurgical test work by study phase.

METALLURGY					
Process test work	**Detail**	**Description**	**SS**	**PFS**	**FS**
Sample characterisation	1. Assay	Often assays are available for initial sample characterisation.		✓	✓
	2. Mineralogy (XRD)	Hand-held XRD data may be available from geological investigations, calibrated to a subset of analytical XRD analyses.		✓	✓
	3. Mineral liberation			✓	✓
	4. Mineral association			✓	✓

TABLE 5.2 CONT...

Process test work	Detail	Description	SS	PFS	FS
Comminution	1. Bond ball mill work index (BBWi)	A selection of tests may be undertaken. Not all are required, and often geographical location may inform which tests are undertaken. Generally start with BBWi and BRWi initially, expanding as applicable. JK parameter test work has a large sample requirement. HIT test work may be undertaken in later study phases once the comminution characteristics are well understood. Regrind test work, including Isa mill test work, is generally undertaken in the Feasibility Study, but may commence in the Pre-Feasibility Study.	✓	✓	✓
	2. Bond rod mill work index (BRWi)		✓	✓	✓
	3. Bond crushing work index (BCWi)			✓	✓
	4. Abrasion index			✓	✓
	5. JK parameters (aka SAG Mill comminution [SMC] test) (Axb, DWi)			✓	✓
	6. AG mill			✓	✓
	7. Starkey			✓	✓
	8. Isa mill				✓
	9. Regrind				✓
	10. High pressure grinding rolls				✓
	11. Rapid HIT			✓	✓
	12. Ore sorting			✓	✓
	13. Unconfined compressive strength				✓
	14. Materials handling test work (eg TUNRA)				✓
	15. Dry grinding				✓
Heap leach / *in situ* leach / vat leach	1. Agglomeration	Assessing conditions including, for example and where applicable: grind, reagent types / additions, temperature, residence time, modifiers.		✓	✓
	2. Bottle rolls		✓	✓	✓
	3. Columns			✓	✓
	4. Aquifer permeability			✓	✓
Tank leach / pressure leach	1. Bottle rolls	Where the process is integral to the material recovery, test work is anticipated in all study phases. Conditions may include grind, reagent types / additions, temperature, residence time, modifiers, oxygen update.	✓	✓	✓
	2. Agitated beaker tests		✓	✓	✓
	3. Pressure vessel tests		✓	✓	✓
Settling	1. Kynch tests	Kynch tests may be undertaken, particularly where there is insufficient sample available to undertake the test work. Dynamic test work may be undertaken by an independent laboratory or equipment provider, depending upon the study phase and requirements for process guarantees. Determine rise rate, settling rate and overflow clarity. Rheological test work is generally undertaken with thickener underflow samples at a variety of densities.		✓	
	2. Flocculant screening			✓	✓
	3. Dynamic test work				✓
	4. Rheology				✓
Filtration / transport moisture limits	1. Vacuum	Filtration test work may require flocculant addition, but not always. Transport moisture limit test work is generally only required in the Feasibility Study, unless for a new product, in which case test work may be undertaken in the Pre-Feasibility Study as well.		✓	✓
	2. Pressure			✓	✓
	3. Flocculant screening				✓
	4. Transport moisture limits				✓

TABLE 5.2 CONT...

Process test work	Detail	Description	SS	PFS	FS
Flotation	1. Flash/coarse particle	Consider grind, residence time, reagent combinations, pH.		✓	✓
	2. Rougher	Pilot plant test work generally undertaken for new processes and to generate products for downstream test work.	✓	✓	✓
	3. Cleaner		✓	✓	✓
	4. Other (eg Jameson)				✓
	5. Batch		✓	✓	✓
	6. Locked cycle test work			✓	✓
	7. Pilot plant				✓
Gravity	1. Nelson/Falcon			✓	✓
	2. Dense media separation		✓	✓	✓
	3. Magnetic separation		✓		✓
Solvent extraction	1. Shakeout			✓	✓
	2. Pilot plant			✓	✓
Precipitation	1. Agitated beaker tests			✓	✓
Tailings	1. Static geochemical tests	Generally undertaken with composites that particularly represent the mining payback years, if possible.		✓	✓
	2. Kinetic geochemical tests			✓	✓
	3. Static settling and consolidation tests			✓	✓
	4. Geotechnical tests			✓	✓
Corrosion	1. Pot tests	Coupon test work, slurry/liquor and vapour space conditions where public domain information not available. Variable composition / pH / oxidation reduction potential as applicable.			✓
Gold recovery	1. Carbon-in-leach		✓	✓	✓
	2. Carbon-in-pulp				✓
	3. Carbon activity				✓
	4. Equilibrium carbon loading				✓
	5. Cyanide destruction				✓
Pyrometallurgy	1. Pelletisation	Undertaken in the Scoping Study and Pre-Feasibility Study. Laboratory test work generally considered empirical and indicative only. Pilot plant test work generally undertaken for the Feasibility Study.		✓	✓
	2. Calcination		✓	✓	✓
	3. Roasting		✓	✓	✓
	4. Smelter		✓	✓	✓
	5. Refining		✓	✓	✓
Pipelines	1. Rheology	For long pipelines, specific test work may be required to ensure the pipeline and pumping systems are fit for purpose.			✓
	2. Slope settling tests				✓
	3. Corrosion (see Corrosion)				✓
Water	1. Assays	For all service requirements.		✓	✓
Reagents	1. Assays	Depending on the process, test work with reagents may need to be undertaken to determine potential impurities and reactivity.		✓	✓
	2. Mineralogy (XRD)				✓

SS = Scoping Study; PFS = Pre-Feasibility Study; FS = Feasibility Study; XRD = X-ray diffraction; Axb = ore breakage parameter; DWi = drop weight index; HIT = hardness index tester.

SAMPLE PROVENANCE

At any phase in the study process it is imperative to ensure that clear and accurate records are kept, which document:

- Sample form including single point locations such as test pits or drill core sections, as well as composites. Single point location and composite samples are generally prepared on the basis described below. Single point locations enable test work information to be incorporated within the geology model and output in mine plan optimisation investigations. Composites may be required for the following test work:
  - Often larger samples are required for pilot plant test work, and composites are generally prepared to support this type of test work.
  - In addition, composites may be prepared to enable parameterisation test work to be undertaken due to the larger sample requirements.
  - A third basis for composite preparation is to replicate an average feed that a process plant might see based on the mine plan. This basis is often flawed as mine plans constantly change through study phases. Instead, composites prepared to confirm or explore additivity often provide the same information independent of any mine plans.
- Sample selection basis, ensuring the Resource characteristics are understood to the level commensurate with the study phase and Resource complexity. Characteristics may include elemental analysis, mineralogical analysis, rock quality data, geological structure and fault identification. Within the areas of elemental and mineralogical composition, the metallurgist may consider deleterious elements where applicable. A discussion should be included as to why the number of samples selected is sufficient for the study phase.
- Sampling process and storage (from origin to intermediate and final products). Some drill core samples may be stored without any special conditions, whereas others require inert atmospheres and refrigeration, for example. The basis for sample selection will be informed by the deposit type, for example sulfide deposits may require samples to be refrigerated. These requirements have a cost – both in terms of sample viability and condition requirements.
- Physical sample location coordinates and descriptors (eg drill hole identifier and position).
- Maps in 2D or 3D, relating key elemental, mineralogical and geological characteristics to the Resource and sample locations.
- Any other key information.

TEST WORK

Historical test work

There may have been historical test work undertaken that could contribute valuable data/ information to any study phase.

This test work would be expected to be either summarised if no longer relevant to the definition of the project being investigated, or described in further detail where the information is applicable to the definition of the project in its current form.

Test work plan

The metallurgical test work plan may be a document specific to the study phase or a living document, depending on the project. It is used to set the scene and assist in scoping and budgeting test work, as well as inform to the study phase any specific requirements for metallurgy.

For a project that is specific to a study phase, there are common areas that should be addressed, and these include:

- *Introduction* – Setting the scene for the project phase
  - *Objectives* – What are the objectives of the test work plan?
  - *Assumptions and constraints* – Project specific
  - *Approach* – Phase specific, activities and type of test work proposed
  - *Project schedule* – Where known.
- *Material characteristics, sample selection and preparation* – Sample selection basis as described above, with consideration to the test work to be undertaken with each sample. Generally include a test work matrix, particularly from the Pre-Feasibility Study, identifying what test work will be undertaken with each sample. Mineral characteristics will inform which type of test work should be planned to be undertaken.
- *Test work plan* – This may cover multiple sections in the document and is expected to address:
  - Laboratory and material flow structure between laboratories, where applicable.
  - Processes to be tested, and an if not/why not explanation for processes not being tested where material is available. For example, where a process is well known and there are multiple operations with minimal variation between facilities, piloting may not be proposed.
  - Test work process definition for all processes to be tested:
    - In a Scoping Study, industry standard parameters may be applied in test work for key processes only. This may include reagent selection, residence time and physical sizing, for example. Test work at this phase may be focused on proving the process concept only.
    - In a Pre-Feasibility Study, a range of parameters may be tested in order to optimise conditions by metallurgical domains, and variability test work undertaken to demonstrate those conditions. Additional processes may also be tested at this time to support flow sheet option investigations and other processes within a flow sheet. Variability test work may comprise two types of test work: batch, and semi-continuous or continuous test work. Some parts of the test work may be described as confirmatory test work.
    - Test work in a Feasibility Study will generally comprise further variability test work for key processes as well as limited test work for remaining processes as required.
    - Types of test work that may be undertaken according to phase are referenced in Table 5.2.
  - Flow sheets for processes to be tested including batch and semi-continuous or continuous test work.
  - Simulation/modelling as required to set test work conditions including, for example, solution chemistry, operating densities and solids compositions. This information may also be used to prepare synthetic solutions as required for test work.

  - Reporting criteria including photos, video, conditions, sampling, mineralogical investigations, assaying and quality assurance / quality control defined for the test work procedures.
- *Process development* – Describe what the test work will inform or enable, including for example:
  - Orebody definition for metallurgical purposes, also known as geometallurgical characterisation. This may include recovery and predictive relationships, etc referred to below.
  - Flow sheet definition.
  - Process design criteria.
  - Mass balance development.
  - Equipment sizing.
  - Predictive recovery relationships for specific processes/overall process.
  - Predictive relationships for key equipment sizing parameters including, for example, comminution, leach and flotation residence times and reagent requirements.

The test work plan would normally be incorporated in an appendix within the study report, with a summary incorporated into the study report metallurgy section.

The test work plan will be developed to the extent that reflects the requirements for the study phase. As described in Table 5.1, only key processes may be tested during a Scoping Study. Test work to support option investigations may be undertaken in Scoping and Pre-Feasibility phases.

The metallurgical test work plan may be expanded to include environmental and geotechnical test work, amongst other areas.

While the test work plan may include a schedule, depending upon the outcomes as the test work progresses, metallurgical test work may take longer than planned. This happens often enough that planning for this within the study phase is essential. This may include, for example, planning to iterate the engineering evaluation late within the study phase, or risk assessing metallurgical test work plan changes to move test work from one study phase to the next.

Test work results/reporting

There are essentially two types of reporting: reports from the test work facilities and test work interpretation.

The former is addressed here. The latter is addressed in the following section.

Test work reports are considered primary records for the metallurgy investigations undertaken. These reports document the journey, and generally include:

- sample identification and blend development
- scope of work
- test work procedures
- test work results, including detailed test reporting
- test work interpretation, where applicable
- documentation of any particular issues encountered.

Regardless of the study phase, all test work reports should contain these elements. Requiring only spreadsheet files of results as deliverables is not good practice for reasons, including:

benchmarking of test work procedures with other laboratories, comparing test work results across laboratories, troubleshooting test work results where test work results are not as expected, etc.

Test work interpretation

Test work interpretation informs the flow sheet design and scale up from bench scale or pilot plant test work to commercial operation, value element recovery and reagent requirements, for example.

Models/geometallurgy

Test work enables the development of models, often relating elements, minerals and other criteria to:

- recovery/rejection for value and deleterious elements and/or minerals
- reagent/consumable additions
- comminution parameters
- other information.

The relationship between geological information (assays, mineralogy, other criteria) and metallurgical information is often referred to as geometallurgy. Historically this is also known as mine to mill.

In Scoping Studies generally only preliminary test work information is available, and this information is typically insufficient to develop models that may then be input into the Resource model with determination of relevant values by block, or for mine plan optimisations.

Ultimately these models are updated through study phases as increasing amounts of relevant information become available. The form of the models may also change as knowledge of the Resource improves, and confidence is increased.

Scale up

There are many resources available that provide information on scale-up factors, including, for example, the Society for Mining, Metallurgy and Exploration proceedings.

The application of scale-up parameters may be based on industry standards, where possible, or may be an assumption where a new process may be considered in a Scoping Study.

As further test work information becomes available, scale-up parameters may be updated to better reflect the technology application. For new processes, piloting may be required in order to demonstrate the process and enable scale-up parameters to be defined more appropriately.

Other information

Other information that may be determined from metallurgical test work may include:

- equipment type
- equipment sizing parameters
- residence time
- reagent requirements.

FLOW SHEET DEVELOPMENT

During the Scoping Study phase multiple process options may be assessed at a preliminary financial level, with a single or multiple option(s) selected and recommendations for future investigations at the end of the study. Flow sheets will be developed at a block flow level and may inform and/or be informed by test work.

Future investigations may include metallurgical test work to confirm assumptions in the Scoping Study phase. This will increase confidence in flow sheet selection and flow sheet parameter optimisation. These test work investigations may identify options that are then revisited in the Pre-Feasibility Study phase.

In a Pre-Feasibility Study, the flow sheets will be developed further, showing all equipment listed in the equipment list.

Assessment criteria are addressed later in this handbook (see Chapter 10); however, elements other than financial metrics may also be assessed, including, for example, social and environmental impacts.

By the end of a Pre-Feasibility Study the flow sheet that will be carried forwards into the Feasibility Study phase should be finalised because the Feasibility Study phase is then focused on finalising designs to enable a financial investment decision to be made by the board.

PROCESS DESIGN CRITERIA / MASS BALANCE

As a project progresses through the study phases the process design criteria and mass balance development will continue, with increasing detail and confidence in the parameters defined.

Process design criteria

The purpose of process design criteria is to define the process design and enable sizing of equipment, and to define inputs into the mass balance, which may also be used to size equipment.

Key information that is input to structure and populate process design criteria may include:

- Work breakdown structure, or key process, reagents and utilities areas. The level of detail will reflect the availability of information and level of information required to meet test work and engineering, and estimate accuracy requirements.
- Units, abbreviations.
- Engineering design criteria. These criteria are normally reported separate to the process design criteria from Pre-Feasibility and will generally include:
    - location
    - annual and seasonal weather parameters, as well as design event parameters (most engineers will be familiar with temperature parameters for equipment design, and rainfall parameters for dam designs)
    - earthquake zone and design basis
    - equipment including capacity factors, key international standards
    - materials selection summary (this is often represented on process flow diagrams – PFDs)
    - piping velocity.
- Information source and referencing. Cross-referencing within the process design criteria.

- Project-specific availability and utilisation, and project life. Availability and utilisation may vary in accordance with process area requirements, for example utilities systems are generally required to have the highest availability requirement specified as these systems are often required to be operating when other process areas are not operating.
- Operating schedule, including hours, days and months, for example.
- Design inputs including:
  - process plant throughput or design production target
  - process source including tonnes, grade, recovery (recovery may be represented by values or an equation)
  - product quality, including, for example, grade, quality (size fraction, loss on ignition, etc as relevant for the process and marketing requirements), deleterious elements
  - depending upon the study phase, single point design inputs and/or an operating range and maximum may be defined.
- Key equipment process design criteria in accordance with the flow sheet and work breakdown structure.

The process design criteria will typically define the design envelope for key equipment. The design envelope may not be specified in a Scoping Study, and implications for equipment selection may not be understood during a Scoping Study. For example, equipment may not operate well over the required design envelope, which may lead to a requirement for multiple equipment items within a process area or full process lines.

What is important, and often happens, is that information is often repeated in multiple documents, and these documents then need to be maintained. In order of priority, the process design criteria document is considered a primary document. Secondary documents may include, for example:

- mass and energy balances
- water balances
- equipment sizing calculations
- reports.

Mass and energy balance

Mass and energy balances are generally developed through each study phase. The mass balance may include the water balance. A separate water balance is often developed for the overall project area, which is generally based on project area weather information, in order to quantify maximum and minimum flows, and to understand the average flows.

During a Scoping Study the mass balance may be very simplified, perhaps to the level of solids flows only, with key elements. A very high-level, static water balance may be developed.

In a Pre-Feasibility Study, the mass and energy balance would be extended to include the entire project. It may include the reagent systems and utility systems where they are integrated with the process. This may include, for a pyrometallurgical process for example, an off-gas system raising steam to generate power.

In a Feasibility Study, the mass and energy balance would be extended to include the design envelope proposed for a project. All major processes will be modelled at this phase.

These mass and energy balances would be expected to be used to inform equipment and facility sizing calculations and operating cost estimates.

PROCESS FLOW DIAGRAMS

The PFDs, otherwise referred to as process flow sheets, are developed from a schematic overview in a concept study (block flow diagram) to a series of detailed and annotated schematic drawings as the study process matures towards Feasibility Study completion. The Feasibility Study PFDs should be sufficiently developed to 'freeze' the process design ready for execution. This means that a single processing option has been selected, reviewed and tested, all major equipment has been sized and laid out and there is little scope for change.

Concept study PFDs are typically very high-level drawings with major equipment steps nominated and major process streams identified. A detailed block flow diagram is sometimes used instead of multiple PFDs in the mainstream concept studies.

PFDs for a Pre-Feasibility or Feasibility Study should include as a minimum:

- schematic representations of the numbered mechanical equipment and plate work items (tanks, hoppers, chutes, etc) that link to the mechanical equipment list and plate work list
- identified and numbered vendor packages where process information is provided by others and included in engineering documentation in a separate format
- numbered streamflows with direction of flow indicated (numbers link to the mass, water, heat balances)
- title block (dated and signed).

PFDs are often annotated with equipment details, mass balance tables and other information. This can be informative but may result in duplication and issues with version controls because the source data may be in other engineering documents.

GAP ANALYSIS

At the end of each study phase a forward workplan should be completed, including recommendations for test work to improve confidence in decisions.

The gap analysis will include follow-up work from the phase for which it is undertaken. In addition, the gap analysis will address the forward workplan for the next project phase.

FURTHER READING

Davoodi, M, 2017. Metallurgical testwork for feasibility studies: what bankers and decision makers should know [online], 26 October 2017. Available from: <https://www.linkedin.com/pulse/metallurgical-testwork-feasibility-studies-what-know-davoodi-p-eng-/>.

Investing News Network, 2018. Why is metallurgical test work important? [online], 22 July 2018. Available from: <https://southernsilverexploration.com/site/assets/files/3982/why_is_metallurgy_test_work_important.pdf>.

Mular, A L, Halbe, D N and Barratt, D J, 2002. *Mineral Processing Plant Design, Practice, and Control,* Volumes 1 & 2 (or latest version) (Society for Mining, Metallurgy, and Exploration, Inc: Littleton).

PROCESS PLANT AND INFRASTRUCTURE

6

INTRODUCTION

Purpose

As with all areas of a study, the main process plant and infrastructure study foci are determined by managing project risk, in this case risk pertaining to process selection, design and operation, thereby maximising project value.

Chapter 5 outlined the approach to metallurgical development, geometallurgy and engineering design test work. This information feeds into the consideration of flow sheet options, flow sheet design and the development of engineering data that support estimating and economic evaluation. This chapter summarises:

- the plant design process and deliverables and reporting outcomes for each study phase
- the types of study documentation across the engineering disciplines: process, earthworks, civil, structural, mechanical, piping, electrical, control and instrumentation, and architectural.

Maximum value and minimum risk are attained by considering a broad range of processing options in the early screening phases, typically in a Scoping Study. This screening is informed by test work and other risk management practices as the process plant design proceeds through the Pre-Feasibility Study phase (to select the likely highest value process plant options) and then through the Feasibility Study phase (to provide validation of project value using the best process plant outcome).

The determination of the best process plant outcome involves identifying the option with the highest net returns. It is therefore an exercise in optimising the mix of throughput, recovery, operating costs, realisation costs and capital cost. It is critical that these elements are considered and optimised at each study phase.

Types of process plants and infrastructure

The process plant upgrades the feed material to a concentrate, salt, metal or other marketable product. The specifics of a process plant design are determined by geometallurgical characteristics of the feed material and market acceptance of the product. Markets for bulk materials, base and precious metals tend to be well understood. Markets for industrial minerals/products are less well defined and more specific to each end user / contract.

Process plants for mineral processing and the metallurgical industry cover a range of scopes, including (but not limited by) one, or a combination of the following:

- Comminution processes targeting reduction of the feed particles size. Comminution typically forms a part of any processing circuit as particle size minimisation is often critical for value recovery.
- Bulk leach, eg dump leach, vat leach, heap leach. These plants are typically followed by hydrometallurgical metal recovery processes, for example carbon adsorption for gold recovery or solvent extraction / ion exchange for copper or uranium recovery.
- Mineral separation / concentration / purification – including gravity and flotation plants as well as specific techniques (magnetic separations / ore sorting / solvent extraction / ion exchange / membrane separation) producing mineral concentrates, loaded liquors and waste tailings streams.
- Hydrometallurgical plants producing salts, precipitates or metal products from solutions.
- Pyrometallurgical processes treating ground ore or concentrates or other intermediates to produce metal or matte products, or as a pre-treatment for hydrometallurgical processing.
- Refineries treating intermediate products to produce high-purity metal products. These can be electrowinning and electrorefining, precipitation and calcination plants.
- Infrastructure including power, water, tailings, reagents and other utilities to enable processing as shown in Table 6.1.

TABLE 6.1

Infrastructure areas

INFRASTRUCTURE		
Area	**Subareas**	**Comment**
Power	Main substation Power station	Where a power station is considered, assess fuel options, renewables and path to decarbonisation.
Water	Raw water Treated water Potable water Cooling water Process water Recycled water	
Roads	In plant roads Site access roads Construction access roads	Consider road improvements/development to enable construction access where different from site and in plant road access.
Rail	Rail line Rail engines Wagons Siding	Consider level of automation applicable to project.
Port	Existing New	Consider requirements for construction, operations; customs; draft/size of ships; cranage.
Airstrip/helipad	Existing New	Consider requirements for operation, eg emergency access, fly-in, fly-out, construction, product transport.
Air systems	Plant Instrument Process	Consider whether all air systems to be dry, based on operating environment. Consider process: pressure requirements, materials of construction.

TABLE 6.1 CONT...

Area	Subareas	Comment
Workshop	Construction On-site Off-site	Consider requirements during construction. Depending on location, may require on and off-site facilities.
Warehouse	Construction On-site Off-site	Consider requirements during construction. Depending on location, may require on and off-site facilities.
Buildings	Construction On-site Off-site	Consider requirements during construction. Depending on location, may require on and off-site facilities.
Communications	Radio Process control and instrumentation Mobile Satellite	
Accommodation	Camp Local town	Consider requirements during construction. Depending on location, may require on and off-site facilities.
Conveyors and pipelines	Water Fuel Reagents Ore Products	Generally connecting sites, eg mine to process plant; port to process plant; process plant to tailings storage facility.
Waste	Packaging Glass Oil Rubber Metal Process waste Sewage	Consider requirements during construction. Depending on location, may require on and off-site facilities. Consider how waste will be managed in operations.

The intensity of design for a process plant and infrastructure at each study phase is determined by what is being evaluated and the purpose of such evaluation. The early phase of the design involves the benchmarking of similar processes, sourcing test work data, identification and comparison of processing options and the management of risks, together with the consideration of other project component interactions, to determine the likely highest value process plant outcome.

Study context

Process plant and infrastructure design evolves during the study phases, from Scoping to Feasibility, as more information becomes available. It is initially driven by process engineering activities such as test work interpretation and flow sheet development. As these activities mature and the preferred options emerge, process plant design requires support of other engineering disciplines. Depending on the nature of the selected process, a large-scale test work campaign and/or piloting may be necessary to obtain sufficient information for a process plant design and to de-risk the project. Thus, there is an interplay between process plant design, metallurgy, geology and mining study components that continues through the study sequence. The metallurgical and engineering testing processes are used to reduce project risk and improve confidence in the selected process plant design.

Unless a study is a preliminary desk-top evaluation of a hypothetical process, it should not commence without at least rudimentary test work data pertaining to the material to be processed. A conceptual process plant design is the appropriate vehicle to assess alternative processing options and nominate desired test work programs at a Scoping Study phase. The purpose of a Scoping Study is to identify, nominate and study potential process alternatives and recommend viable options to be brought forward to a Pre-Feasibility Study. The broad-based scope of a Scoping Study is critical to selecting the highest value outcome for the project. A Scoping Study based on only a single preferred option is likely to require rework at a later phase when another potentially feasible and cost-saving alternative is identified or suggested by an independent experienced reviewer.

Engineering context

- Scoping Studies are based on benchmark data and 'factors' derived from previous projects. Pertinent test work data is required to ensure that this approach successfully compares to reference projects.
- Pre-Feasibility Studies use preliminary vendor data (typically library sourced) and bulk material quantity estimates to achieve the required level of accuracy of the project at a relatively low level of precision (percentage of engineering and design completion). Process plant options are further developed during a Pre-Feasibility Study often in parallel to, or preceded by, the execution of recommended test work activities.
- Feasibility Study engineering and design is sufficient to highlight major material risks and to achieve funding of a project.

Where the engineer does not have the required and relevant experience to evaluate the range of options at any of the above phases, there are options to achieve the required accuracy:

- subcontract the scope to an experienced engineering company or consultant
- rely on package vendors (eg a roaster and/or acid plant package) – this should be done with caution as the scope interfaces need to be clearly defined in order to avoid gaps in scope and cost
- engineer and design to a higher level of precision – typically only used for novel process options noting that there are strong links between process novelty and commercial risk that are not always adequately managed by design precision.

A combination of the above may be necessary as the vendor package interfaces need extensive evaluation and analysis with resultant higher levels of precision to provide the 'glue' that results in effective project definition. In the absence of relevant benchmark information and experience, the engineering and design process is likely to be an expensive and long process that commences with extensive metallurgical testing, evaluation of materials of construction, coordination with vendors and evaluation of potential designs at much higher levels of detail and precision than would otherwise be required.

Hence, engineering context is critical for determining the approach used to establish project feasibility. Where an engineer has extensive experience in the delivery and execution of similar process plants in similar geographical areas, the engineer can direct effort at maximising accuracy based on execution-related matters. Where an engineer is developing new or novel processes or relying on vendor input, the focus is on understanding and defining the detail of the interfaces, necessitating additional detail and precision in the engineering of the interfaces.

PROJECT SCOPE AND GENERAL PLANT DESIGN REQUIREMENTS

General

Process plants and infrastructure are comprised of:

- equipment that may be manufactured by original vendors (original equipment manufacturers) or by others
- vendor-supplied packaged plants
- instruments, valves and other control system elements
- bulk materials such as piping, structural steel, platework, electrical and communication cables
- buildings (bespoke and/or packaged)
- all with supporting civil works (concrete) and earthworks (bulk and detailed).

The engineering design of process plants includes the production of a series of engineering deliverables – that is documents and drawings – for each of these components. The level of development for such deliverables varies depending on the phase of a study and maturity of process development.

To manage the production of deliverables, the process plant scope is typically split by 'work breakdown structure' (WBS) into the various process plant facilities and into 'disciplines' based on engineering discipline. The design and formulation of this matrix is a critical step that needs to be initiated early in the study sequence and updated as the project develops. Failure to correctly define the WBS by discipline structure will result in difficulties for reviewers and misreporting of costs.

The scope and general requirements for the process plant component of a study become more detailed as the project progresses from Scoping Study, through Pre-Feasibility Study to Feasibility Study. Table 6.2 summarises the high-level key requirements for the process plant and infrastructure. Specific engineering requirements are discussed later in this chapter.

TABLE 6.2

Summary of process plant and infrastructure study requirements.

PROCESS PLANT AND INFRASTRUCTURE			
Information	**Scoping**	**Pre-Feasibility**	**Feasibility**
Project location	Preliminary – options identified	Required – preferred option selected	Required
Feed source	Required	Required	Required
Product specification	Required	Required	Required
Capacity or throughput	Preliminary – options identified	Required	Required
Scope of work and battery limits	Preliminary – options identified	Required	Required
Work breakdown structure	Required (Level 2)	Required (Level 3)	Required (Level 4)
Engineering and design inputs	Benchmarked	See Table 6.3	See Table 6.3
Plant production profile and schedule	Outline	Required	Required

Scope interfaces

The process plant scope of work interfaces with numerous areas of the study, including but not limited to:

- *geology* – ore characteristics, including mineralogy and geometallurgy (Chapter 3)
- *mine or feed supply* – process plant feed supply rate and physical characteristics (Chapter 4)
- *test work and flow sheet development* – typically conducted in stages aligned with study development (Chapter 5)
- *tailings or waste product storage* – long-term waste storage approach (Chapter 7)
- *environmental, social and governance activities* – the potential impact of the plant on external parties (Chapter 2)
- *project management* – including estimating, logistics, execution planning and scheduling/ sequencing (discussed in Chapter 8).

In addition, the process plant scope of services interfaces with other aspects of the project, such as:

- business accounting systems
- asset/material management
- tender documentation
- operational readiness and operations planning
- owner's scope.

All interfaces need to be mapped and understood with sufficient definition to suit the project. 'Sufficient definition' needs to be consistent with the project risk and contingency allowances for unknown factors, whether they are technical, geographic, political, social, environmental, capability, capacity or cost related.

All assumptions made in relation to the process plant scope need to be clearly stated, preferably with a justification of the relevance of the assumption.

PROCESS PLANT AND INFRASTRCUTURE ENGINEERING

Summary work program

The process plant and infrastructure (PP&I) engineering and design is defined in a series of design criteria documents and supporting deliverables. Table 6.3 summarises the PP&I engineering deliverable for each phase of the study. During the Pre-Feasibility and Feasibility phases the deliverables are further grouped by WBS.

Process plant and infrastructure general

Project and engineering management

At the Scoping Study phase, engineering management is often performed by a process lead, acting as a process plant engineering manager. As the study matures, a dedicated project engineering manager takes over the management of the process plant engineering as a part of their duties.

TABLE 6.3

Summary of process plant and infrastructure engineering and design deliverables.

PROCESS PLANT AND INFRASTRUCTURE				
Discipline	**Activity/deliverable**	**SS**	**PFS**	**FS**
General process plant	Engineering management and reporting		✓ (for complex/ large study)	✓
	Discipline management and reporting			✓
	Process plant study chapter	✓	✓	✓
	Work breakdown structure	Level 1	✓	✓
	Process plant basis of design		✓	✓
	Reagents and utilities list		✓	✓
	Engineering deliverables list		✓	✓
Process design and drafting	Process design criteria	✓	✓	✓
	Block flow diagram – incl. options	✓	✓	✓
	Process flow diagrams		✓	✓
	Mass/energy balance	✓	✓	✓
	Process description	✓	✓	✓
	Options description	✓	✓	
	Engineering/vendor test work reports		✓	✓
	Large-scale test work / piloting reports		✓	✓
	Plant capacity analysis		✓	✓
	Piping and instrumentation diagrams – process input			Depends on plant type
	Process control philosophy		✓	✓
	Operating cost estimate calculations	✓	✓	✓
	Reagent first fill calculations			✓
Mechanical design and drafting	Mechanical design criteria			✓
	Mechanical equipment list	✓	✓	✓
	Equipment data sheets		✓ (long lead items)	✓
	Equipment sizing calculations		✓ (long lead items)	✓
	Materials of construction list		Required where MOC is critical	✓
	Equipment specifications			✓
	Mechanical general arrangement drawings		Preliminary	✓
	Overall plant layout		✓	✓
	Platework MTOs		Preliminary	✓
	Mobile equipment list		Preliminary	✓
	Spares list		Preliminary	✓
	Plant 3D model input (mechanical)		Preliminary	✓
	Sketches and area layouts – plan and elevations		✓	✓

TABLE 6.3 CONT...

Discipline	Activity/deliverable	SS	PFS	FS
Mechanical design and drafting (*continued*)	Performance specifications for packaged plants and utilities		✓	✓
	Heating, ventilation and air conditioning requirements for detailed design		Preliminary	✓
Piping design and drafting	Piping design criteria		Preliminary	✓
	Piping and instrumentation diagrams – piping selection and sizing			Depends on plant type
	Piping isometric drawings			If required
	Piping and valve specifications			✓
	Piping material take-offs (MTOs)		Large bore only	✓
	Valve lists			✓
	Piping line list			✓
Electrical design and drafting	Electrical design criteria			✓
	Single line diagram		HV only	✓
	Electrical equipment list		✓	✓
	Electrical specifications			✓
	Electrical load list		Preliminary	✓
	Electrical, instrumentation and control designs for all process facilities and on-site infrastructure			✓
	High voltage and medium voltage electrical design			✓
	Security and CCTV requirements			✓
	Cable schedule			✓
	3D modelling of major electric equipment and cable routes			✓
	Power distribution general arrangement drawings (includes buried and overhead)			✓
	Input to the plant layout development and incorporating major electrical		✓	✓
	Electrical material take-off			✓
Instrumentation and controls design and drafting	Instrumentation design criteria			✓
	Control systems diagrams			✓
	Control system and communication block design			✓
	Instrument list			✓
Geotechnical	Maps and survey data		✓	✓
	Soil and foundation geotech		✓	✓
	Geotechnical drilling/investigation reports		✓	✓
Earthworks	Earthworks design criteria			✓
	Bulk earthworks engineering specifications			✓
	Bulk earthworks, site access and on-site roads and drainage designs in 2D			✓

TABLE 6.3 CONT...

Discipline	Activity/deliverable	SS	PFS	FS
Earthworks (*continued*)	Importing and validating the detailed digital terrain model with XYZ coordinates and ground survey data			✓
	On-site traffic movement plans		✓	✓
	Earthworks MTOs		Preliminary	✓
Civil, structural and architectural	Civil and structural design criteria			✓
	Civil layouts on topography		✓	✓
	Civil drawings			✓
	Foundation design		Preliminary	✓
	Civil MTOs		Preliminary	✓
	Plant 3D model – structural		Preliminary	✓
	Structural general arrangements			✓
	Structural MTOs		Preliminary	✓
	Architectural design criteria			✓
	Architectural specifications for site buildings			✓
	Architectural 2D layouts for on-site buildings			✓
Risk and hazards	Process risk assessment	✓	✓	✓
	Hazard identification		✓	
	Hazard and operability study			✓
	Hazardous area diagrams			✓

Work breakdown structure

The WBS defines the scope and structure of the project and is developed for the overall project (refer to Table 6.4). A logical work breakdown structure is critical to defining and understanding the structure of the following:

- process plant areas
- potential work packages
- capital cost estimate.

The process plant is typically a Level 1 component of the WBS, along with mining, site infrastructure, port, transport pipelines and other major components.

The process plant is then split into Level 2 components, for example ore handling, concentrator, flux mixing, smelting, leaching, solid/liquid separation, purification, electrowinning, etc, depending on the nature of the process plant.

Level 3 components are then a further breakdown of each Level 2, and Level 4 is typically at the unit process level, for example cathode stripping or concentrate filtration.

It is important to link the WBS in the study phase to the requirements for contract packaging, commissioning and logical execution of the project. It is also useful for the WBS to represent elements that are able to be benchmarked against the engineer's database of materials and costs for estimate input, estimate validation and cross-checking, depending on the study phase.

TABLE 6.4
Example work breakdown structure (WBS).

WBS	DESCRIPTION	WBS SUBAREAS
1000	Mining	Mine 1, 2, etc; Mine infrastructure; Stockpiles; ROM
2000	Processing section 1	Crushing; Grinding; Flotation; Settling and Filtration; Carbon-in-Pulp/Leach; Cyanide Destruction; Gold Recovery
3000	Processing section 2	Leach and/or Pressure Oxidation; Neutralisation; Counter Current Decantation; Impurity Removal; Metal Refining Smelter; Electric Arc Furnace; Slag Handling; Converters; Product Handling
4000	Tailings	Filtration; Storage Facility; Pipelines; Decant and Seepage management
5000	On-site infrastructure	Power; Water; Water Treatment; Reagents; Consumables; Buildings; Communications; Roads
6000	Off-site infrastructure	Power; Water; Roads; Rail; Buildings
7000	Logistics	Pipelines; Rail; Roads; Warehouse; Product Handling
8000	Indirect costs	Engineering, Procurement and Construction Management; Construction Facilities; Temporary Facilities; Water; Camp
9000	Owners costs	Studies; Environment, Social and Governance; Permitting; Resettlement; Land Acquisition; Owners Team; Operations Readiness; Insurance

Basis of design and design margins

The basis of design document sets out the basis for the overall process plant for each discipline: general project summary, process, mechanical, piping, electrical, structural, civil/ architectural and earthworks. The basis of design reflects the fundamentals of a design, including relevant standards and design allowances. Requirements, and variations, are based on the engineer's experience with specific process plant flow sheets. The level of detail depends on the type of process plant and may refer to a set of design criteria for each of the relevant disciplines.

A Scoping Study does not require a basis of design document for the process plant design. A simple process design criteria document is usually sufficient. A Pre-Feasibility Study requires a basis of design document and this may include non-process discipline design criteria, particularly if the design differs from typical design considerations, eg seismic conditions. A Feasibility Study further develops the basis of design document and typically appends or references discipline design criteria that are relevant to the project.

The basis of design document defines and maps up design performance risk and operational outcomes. These include:

- margins on test work outcomes
- margins in mass, energy and water balances used to select equipment
- margins added by discipline engineers
- margins added by project engineers in the procurement cycle (owner, engineer and/or vendor)
- margins required by clients.

Design margins are necessary to manage project risk. They need to be understood and managed to ensure that the process plant is not overdesigned, and/or that 'over designed capacity' is pragmatically taken into consideration when calculating the project throughput, recovery and revenue as a function of the mine plan or expected feed profile.

All Pre-Feasibility and Feasibility Studies should include a plan that manages design margins across engineering and procurement to ensure the study meets stated targets and avoids unnecessary rework.

Precision and accuracy of engineering data

The level of engineering and design completion for a study is dependent on the risk associated with the project and with the component process plant facilities. The relationship between precision (engineering completion specific to the project) and the accuracy of a study is not directly proportional. Accuracy is a function of the following:

- specific site knowledge
- knowledge related to infrastructure scope and interfaces
- status of environmental, social and governance (ESG) discussions and processes (environmental permitting, alignment with community and government permitting processes)
- knowledge related to the process plant feed material, its variability and process response
- recent experience of consultants and engineers
- inclusion of novel technology.

The project owner's business and project experience is also a key factor, but a junior mining business may be able to achieve a more accurate outcome than a major mining house due to the inherent efficiencies and focus of a smaller business.

As a result of the above, it is not uncommon for studies that have been engineered to a high level of precision resulting in lower levels of inherent accuracy than studies with a lower level of precision. This is particularly the case for process plants where an engineer may have extensive history in project delivery and can use benchmarking of historical data to derive quantities and costs for a facility or the complete plant.

Process design

Process plant engineering starts with process flow diagrams (PFDs) and process design criteria that are developed from the metallurgical and engineering test work discussed in Chapter 5. As the study phases develop, the metallurgy and test work are developed to manage material risks to the project. These risks may be associated with variability in performance due to mineralogy/geology or related to specific aspects of the plant design.

The basis of the process plant design is the 'process package'. The process package consists of the documents required by the other discipling engineers to complete the engineering and design of the facilities to the designated level. The process package consists of the following engineering documents:

- process design criteria
- block flow diagram
- mass balance or mass/energy balance
- process flow diagrams
- preliminary mechanical equipment list
- preliminary instrumented PFDs.

For a Feasibility Study, the process package should be sufficiently developed to support detailed engineering, and require modest rework should the project proceed to execution. It is expected that following the completion of a Feasibility Study the process design is effectively 'frozen' and only supports the discipline's design.

For earlier phase studies, the process package should be sufficient to describe the options considered in the option analysis, and explain the logic used to select a preferred option in sufficient detail for subsequent reviewers.

There is a close link between the process package and mechanical equipment selection, vendor package definition and plant layout. Hence, where processes are novel or where there is limited reference or benchmark data, the process package needs to be more advanced than for process packages where there is a high level of engineering experience. Process plants that consist of substantial vendor packages (for example a roaster and acid plant) require a process package that focuses on specification of the range of vendors of package inputs, outputs and duties rather than the specifics of the equipment within the vendor package.

As a study progresses from a Scoping Study phase and the options are progressively shortlisted and selected, the following deliverables are developed based on the process package documents. These include:

- process description
- process control philosophy
- process plant capacity analysis
- inputs to hazard studies
- inputs into capital (first fills) and operating cost estimate
- equipment data sheets
- technical evaluations of vendor quotations
- review and interpretation of large-scale test work / piloting and applicable vendor test work reports.

PP&I description

The process description is required for an engineer to understand the nature of the process with the aid of the process flow diagrams. The document detail matches the level of flow sheet definition and engineering design.

At Scoping Study level, the process description outlines the major facilities in sufficient detail for the reader to understand the process and the key equipment or packages.

At Feasibility Study level, the process descriptions should include reference to mechanical equipment numbers, valve numbers, plate work numbers and other key data that assists the reader to follow the process flows and process logic.

PP&I design criteria

The process design criteria document tabulates the basis of the process design. The document may present only the inputs to the process design that allow calculations to be made and equipment to be selected, or it may be a more complete process engineering summary document that contains details of the selected equipment as well as the key input data.

The more detailed process engineering summary style requires updating more frequently as calculations change and duplicates can result from information contained in separate process calculation spreadsheets, mass balances and the mechanical equipment list. Engineers who prefer 'one source of truth' tend to opt for the simpler process design criteria that links metallurgical characteristics with the process design inputs used in calculation spreadsheets and mass, heat, water, energy and other balances, particularly if the balances are in modelling packages such as Limn, MetSim or similar products.

Process flow diagrams

The PFDs are developed from a schematic overview in a Scoping Study (block flow diagram) to a series of detailed and annotated flow sheets at Feasibility Study completion. The Feasibility Study PFDs should be close to 'execution ready'.

PFDs for a Pre-Feasibility or Feasibility Study should include as a minimum:

- schematic representations of the mechanical equipment and plate work items (tanks, hoppers, chutes, etc) that link to the mechanical equipment list and plate work list
- indication of vendor packages where process information is provided by others and included in engineering documentation in a separate format
- numbered streamflows with direction of flow indicated (numbers link to the mass, water, heat balances)
- title block (dated and signed).

PFDs are often annotated with equipment details, tables, mass balances and other information. This can be informative but results in duplication as the source data are in other engineering documents.

Process control philosophy

The process control philosophy can be assembled as a descriptive document or in tabular format. The tabular format is more useful for process control engineers, but the descriptive document may be more understandable for the general reader.

The process control philosophy describes how the process is controlled. It is usually initially prepared by a process engineer with input from instrument/control systems engineers for later phase studies and as a precursor to piping and instrumentation diagram (P&ID) development.

The document covers the unit processes, equipment, instruments, valves and other control elements and their variables, set points, limits and control loops.

There should be a focus on personal safety and machine protection when considering safety logic and interlocks. Particular attention is required when considering interfaces between equipment packages or execution packages (eg potential design and construct packages).

Plant capacity analysis

The plant constraint and capacity analysis is often overlooked in studies. Process plants are typically designed for a static condition that considers operational variability (of feed and/or process conditions). This design condition typically represents the average condition plus a design allowance.

Value can sensibility be added to a project by conducting a plant constraint and capacity analysis using the scheduled range of inputs (eg mine schedule or feed schedule). The analysis considers the material characteristics and process outcomes to determine constraints, whether these constraints can be cost-effectively addressed, and the potential gain in project value.

For example, early years of mining may deliver soft ore to the plant that is high grade. The high grade may be due to supergene alteration. If sufficient geometallurgical information is available, throughput optimisation may result in a reduction in cut-off grade and increase in metal units processed and a substantial increase in metal production.

Surge identification may be useful as a part of the plant capacity analysis. This identification addresses perceived bottlenecks in the process and allows for extra capacity be installed, allowing for process interruptions.

Plant capacity analysis is typically performed for Pre-Feasibility and Feasibility Studies.

HAZID, HAZOP and risk assessment

Hazard identification (HAZID) and hazard analysis (HAZAN) studies are required to ensure construction and operation hazards are identified. These early phase activities then feed into hazard operation (HAZOP) studies that document the operability aspects of each hazard and develop elimination of mitigation measures for inclusion in design.

HAZID studies are the minimum requirement of a Pre-Feasibility Study. A HAZOP study requires the development of at least preliminary P&IDs (instrument and control drawings) and are part of the Feasibility Study scope.

The approach to study risks is discussed further in Chapter 11 – Risk.

Engineering design

The engineering effort and the resultant deliverables for each Scoping, Pre-Feasibility or Feasibility Study may vary from project to project. The critical criteria are that the derivation of accuracy needs to be well defined for all plant facilities and the guidelines used by the general market, such as the Association for the Advancement of Cost Engineering guidelines for capital cost estimating, which are guidelines not standards. For example, if the project contains novel technology, higher levels of precision are required to mitigate design risk. Conversely, a project that is similar to a recently executed project should focus on improvements rather than engineering precision.

For a Scoping Study, the disciplines, such as earthworks, civil, structural, platework, piping, electrical and instrumentation are usually estimated as a percentage of identified and major mechanical equipment cost. The percentages are typically based on benchmarking data and previous projects by an experienced engineer.

Following the preferred option identification, the discipline deliverables typically include the items presented in Table 6.3, which are developed to an extent commensurate with the phase and requirements of a specific study.

Mechanical, platework and piping

Mechanical engineering is typically the second most developed discipline after process engineering during the study process. Initial work focuses on the mechanical equipment list and equipment selection. This then flows into engineering associated with materials transport, including conveying, hoppers, pumping, launders and piping to allow the layout of the process plant to commence.

For a Scoping Study, the layout is typically limited to schematic or stylised versions that indicate approximate layout and impact on bulk earthworks and spatial aspects of the project.

In a Pre-Feasibility Study, the equipment needs to be sized and preliminary models fed into computer-aided design or 3D 'cartoons' that allow preliminary take-offs of bulk earthworks, civils and structural steel. Correct equipment selection is critical to defining and optimising layout. It is tempting for a study to allow space in layout due to uncertainty in equipment selection. This will result in overestimation of bulk materials quantities and resulting capital cost. A Pre-Feasibility Study should establish footprint and floor height elevations based on correct slurry hopper and tank dimensions where similar plants have been constructed by the engineer in the past.

Underestimation of wear liner requirements for hoppers and chutes is a common outcome of Pre-Feasibility Studies.

Good plant layout designers are rare. A good designer can reduce plant footprint substantially, thus reducing capital cost.

Civil, structural and architectural

Architectural design usually forms a part of the plant services and infrastructure scope. However, civil and architectural may include buildings associated with the process plant whether protection for climatic conditions or other factors require the process plant facility to be enclosed.

For process buildings, there are two general approaches:

1. the building is a stand-alone structure and outsourced as a separate scope to a building supplier
2. the building is an integral part of the process plant and supports cranes, and pipe rack, or is integrated with equipment.

The stand-alone approach often results in a larger building and is used where construction of the plant is inside the building for environmental/climatic reasons. The larger building is required to cater for design growth and internal structures. The building is on the critical path and allowances are made due to the lack of design detail available at the time the building is put to tender.

The integrated building approach is more common in Australia. The buildings are often not fully enclosed, and protection is only required in specific areas for regular maintenance activities or operational reasons.

Earthworks

Earthworks design requirements are derived from geotechnical conditions, topographical data and plant layout. In hilly locations, the earthworks design balances cut-and-fill, based on estimated material quality. A cost-effective layout will require that process, mechanical and civil structural engineers work together with a designer to consider access, operability, crane requirements, constructability and equipment layout. Getting this right early in the study process will accelerate the study timetable.

Electrical, instrumentation and controls

Electrical, instrumentation and controls are typically factored for a Scoping Study. A Pre-Feasibility Study will start at the electrical load list that is developed from the mechanical equipment list and focuses on high voltage design, substation and motor control centre layout.

A Feasibility Study will develop a full electrical equipment list and cable schedule, together with a list of instruments and control loops. These may then be included in P&IDs, but often a Feasibility Study will be able to be completed using instrumented PFDs that indicate control loops. These documents then lead into development of full P&IDs when piping design is developed post Feasibility. For large bore pipelines or novel processes, P&IDs may be a necessary deliverable from a Feasibility Study.

PP&I PERFORMANCE

Capturing plant performance based on design outcomes is an important part of the study process. This takes two forms:

1. performance warranties
2. performance envelope assessment based on gap analysis.

Process performance is often subject to warranty by the vendors, engineers, constructors and/or project managers. Performance testing is discussed in a paper by Lane *et al* (2007). Performance testing is often used by project owners and financiers to provide protection from, or assessment of, technical aspects of project design, procurement, construction and commissioning. The performance tests can range from comprehensive performance warranties relating to process performance and production in the case of lump sum turnkey projects to less onerous demonstration tests for cost reimbursable contracts and vendor equipment. The requirement for performance testing can be driven by the project financiers, particularly in the case of smaller resource development companies. In such cases, differentiation between owner's risk and contractor's risk can become contentious. At the other end of the spectrum, larger mining houses may choose to heavily influence the engineering and design process with the contractor supplying services on a reimbursable basis. In that case, the performance testing process is often limited to demonstration tests that validate performance for the owner's corporate governance requirements.

Figure 6.1 illustrates the adverse impact of poor performance on shareholder value. Slow project process ramp up reduces revenue and may require further cash to maintain the project during early operation. Any reduction in revenue has an adverse impact on shareholder returns. In extreme cases, slow ramp up can result in significant dilution of the share register or even transfer of company ownership. As a consequence, the process plant designer should always be focused on understanding ore variability, the operability and maintainability of the process plant, associated infrastructure and, in particular, novel or high-risk process elements.

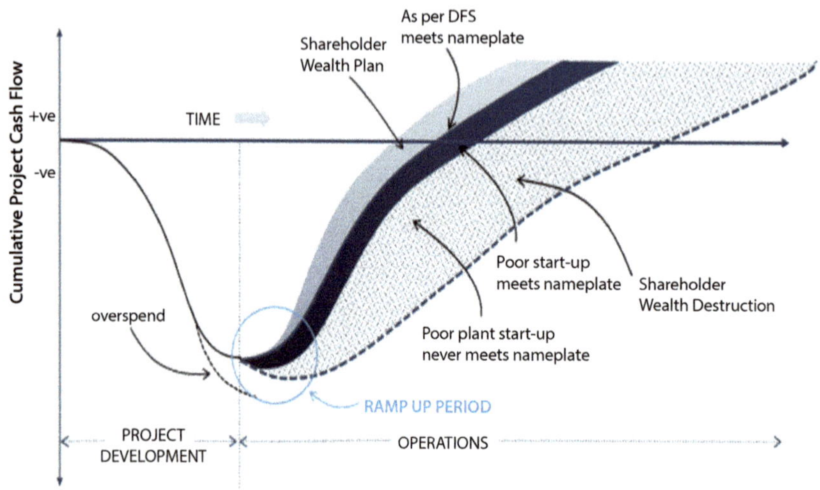

FIG 6.1 – Adverse impact of project performance on shareholder value (Douglas, 2020).

It is important to note that project spending overruns, say 10 per cent, have a lower impact on shareholder wealth than failure (by 10 per cent) to achieve target process plant performance.

McNulty (2014) provides an update on his original paper that covers a range of project types. The 'McNulty curves' can be found in the referenced paper and in many other places in the published literature. The project process 'Types' defined by McNulty cover a range of project start-up outcomes that are fundamentally a function of process selection and engineering design. The descriptions of the 'Types' have been slightly amended:

- *Type 1* – ramp up to design production in less than 12 months requires the use of mature technology, proven similar size equipment, effective pilot testing completion for complex processes, proven benchmark operations, good geometallurgical modelling (or good understanding of the feed source and resultant performance) in place including variability test work, and designers and engineers who know what good looks like.
- *Type 2* – ramp up to 80 per cent of design capacity in 12 months and slow ramp up over 36 months to design typically involves a combination of new technology or prototype equipment, incomplete test work, little or no benchmark operations, severe operating conditions, less developed geometallurgical models based on incomplete variability test work.
- *Type 3* – ramp up to 60 per cent capacity over 12 months. Similar to Type 2 but compounded by lack of knowledge of feed material, inadequate test work, serious design flaws and often 'fast track' such that risks have not been adequately mitigated.
- *Type 4* – ramp up to 40 per cent capacity in 12 months. Similar to Type 3 but with very complex flow sheets that are difficult to pilot. Design has been comprised by optimism and design changes to limit cost, processes are not fully understood and prototype equipment is used.

The definition of project 'Type' and facility 'Type' is an important part of the study reporting process and it requires that the process engineer communicate any areas of risk that may result in the project being anything other than Type 1.

REPORTING

Process plant design is presented as a chapter in the overall study report. Depending on the study (Scoping, Pre-Feasibility or Feasibility) the level of detail increases. Reporting needs to fit the purpose.

The purpose of the process plant chapter of the study report is to explain how the metallurgy and other test work has been interpreted to provide a robust process plant design. A comprehensive description of the metallurgical test work and flow sheet development is presented in Chapter 5. For the purpose of this handbook, the process plant section of the study report should include, as a guide, the information shown in Table 6.5.

CONCLUSIONS

This chapter outlined the requirements for the assessment of process plant and infrastructure design development. The objective of each phase of a study is to provide an accurate (not precise) view of the processes that are applied to extract value. In so doing, the engineer should be focused on increasing project value while maintaining the requisite accuracy.

TABLE 6.5

Process plant inclusions in a study report.

ELEMENT	DESCRIPTION
Overview	
Process description and flow diagram	• Provide a description of the selected process. Nominate any processes that are novel or may present elevated risk. • Provide a simplified flow sheet diagram. If options are studied, structure flow sheets into options and present a basis for evaluation.
Plant design	• Summary format of selected process plant description for Feasibility Study and options as evaluated in the Pre-Feasibility Study. • Tables of equipment type, characteristics and specifications.
Product/materials handling	• Discuss any product or materials handling considerations such as overland conveyors, pipelines, and any operational constraints or requirements.
Energy, water, reagents and process consumables requirements	• Discussion of required reagents, consumables, water requirements (process and potable), including water sources, and energy (power) requirements specifically required for the plant. These can be presented on a life-of-mine average basis but may need to be specific to the payback period for Feasibility Studies where sufficient detail has been developed to support project economics.
Specific requirements	• Are there any specific requirements from smelters/buyers that needed to be considered, and have they been sufficiently addressed?
Process performance	• Process performance as a function of ore variability and mine schedule.
Further work	Next stage of development recommendation: • *Scoping Study* – Summarise options that have been assessed and their respective strengths and weaknesses. Suggest scope of work and services for the Pre-Feasibility Study, including risks and opportunities. • *Pre-Feasibility Study* – Further definition of the selected option in the Soping Study with areas of focus. Specify risks, and particularly any recommended large-scale test work and/or piloting. • *Feasibility Study* – The chapter needs to refer to the program of work required to execute the project.
Appendices	• All engineering data should be provided in appendices. • Project owners should endeavour to recover all data developed by consultants and engineers in native format, where practicable.

In a Scoping Study, the engineer can offer a simplified PP&I design, providing a reasonable view of what the project may be based on, the existing specific knowledge and the engineer's experience with similar projects. All the potential options are typically flagged at this stage and the potentially feasible options are carried over to the Pre-Feasibility Study.

At the Pre-Feasibility Study phase, the engineer needs to consider the full range of potentially feasible options for processing the material and extracting value, but then converge those options on a single likely option in preparation for a Feasibility Study. At this point, all PP&I selections need to be supported by test work, benchmarking, sufficient engineering to define capital and operating costs, and risk assessments.

In the Feasibility Study, up to 20 per cent of overall project discipline engineering may be completed. However, this may be as much as 60 per cent of process engineering and as little as five per cent of piping and electrical engineering. The level of completion is prescribed by the requirement for certified vendor data to proceed in an efficient manner past about 20 per cent of engineering being complete. For a project containing no novel technology, this level (or less) of engineering is sufficient to prescribe costs to the required standard. If novel technology is required, a higher level of process definition is required, occasionally to demonstration plant scale, to achieve an engineered outcome that meets the accuracy requirements. There is no room for unsubstantiated optimism at the completion of the PP&I component of a Feasibility Study.

REFERENCES

Douglas, R, 2020. The art and science of feasibility studies [online], 18 October 2023. Available from: <https://ausenco.com/insights/the-art-and-science-of-feasibility-studies/>.

Lane, G, Davis, M, McLean, E and Fleay, J, 2007. Performance testing – when, what and how?, in *Proceedings Project Evaluation Conference*, pp 197–202 (The Australasian Institute of Mining and Metallurgy: Melbourne).

McNulty, T, 2014. Plant ramp-up profiles – an update with emphasis on process development, in *COM 2014 – Proceedings of the Conference of Metallurgists* (CIM: Montreal).

TAILINGS STORAGE AND WATER MANAGEMENT

7

INTRODUCTION

Tailings storage and water management are interlinked and are essential components of most mining projects. However, poor management of these components can lead to catastrophic failures, loss of life, irrecoverable environmental impacts and social dislocation. No other element of a mining project has the potential to inflict such wide-reaching negative impacts on the surrounding environment and local communities.

High-profile tailings dam failures between 2014 and 2019, including failures in Brazil, Canada and Australia, have led to an increased focus on the engineering design required to safely manage tailings and water for mining projects but also on the governance, sustainability, public accountability and transparency, which is now required from mining companies to successfully finance, permit and develop their projects.

Over the past three years, new standards and guidelines relating to tailings management have been released including ANCOLD (2019a), MAC (2019, 2021a, 2021b), ICMM (2021), GISTM (GTR, 2020) and ICOLD (in press). All these have a common focus in reducing and managing the risk at all phases of project development. This can only be achieved through thorough design, planning and stakeholder engagement well before a project begins development. This requires acquisition of in-depth knowledge of the site, including social setting, and the conditions that could impact the tailings and water management system (ie geology, seismicity, climate, hydrology, hydrogeology, etc). These site characteristics are more often than not obtained as part of the sitewide environmental and social impact studies required for permitting.

To acquire the knowledge base for a successful design of the tailings storage and water management system, a broad mix of professionals such as geologists, geochemists, seismologists, geotechnical engineers, hydrologists, hydrogeologists, social scientists and environmental specialists are required to provide input into the study. But there cannot and should not be a one size fits all checklist of the inputs and deliverables required for each individual study. Each mine site has unique characteristics and therefore requires tailored approaches for investigation and design. All factors that affect the project should be characterised sufficiently to allow the risks that these factors have on the system to be quantified and mitigated to acceptable levels.

This chapter aims to provide an overview of the most common aspects to be considered in the design of the tailings storage and water management systems during the study phases. This chapter does not address the design requirements for raises to tailings storage facilities

(TSFs) that are undertaken during operations. The complex and multidisciplinary nature of tailings and water management prevents this concise chapter from being a comprehensive guide. More detailed and comprehensive references are available and should be employed such as DIIS (2016a and 2016b), ICMM (2021), ICOLD (in press), MAC (2021a) and Morrison (2022).

Tailings storage and water management is a specialised field and studies that cover these aspects should be undertaken by individuals and/or companies who have both experience and specialised knowledge in this area.

STUDY PHASE OBJECTIVES AND ACTIVITIES

The key objectives that should be covered in each tailings and water management study phase are summarised below:

- *Scoping Study* – Develop initial concepts, including identification of potential best available technologies and practices that may be appropriate for the project. The Scoping Study should develop an economically viable site configuration that will have an acceptable risk profile and a feasible closure plan.
- *Pre-Feasibility Study* – Further develop the engineering design of the options identified in the Scoping Study to provide additional certainty on the viability, relative merits and capital and operating costs of the designs. The preferred technology(s) and site(s) will typically be defined during this phase.
- *Feasibility Study* – Advance the design of a single option to support financing, environmental assessments and other regulatory requirements for approval of the project. The closure should be a major consideration in assessing the feasibility of the project.

The activities to be undertaken during the various stages of a tailings storage and water management study are listed in Table 7.1, together with the deliverables and a brief description of the scope within each study phase. These activities are expanded upon in the following sections of this chapter. Although each activity may be addressed at more than one phase, it should be noted that the level of work undertaken in each phase can vary.

TABLE 7.1

Tailings storage and water management study phase guide.

TAILINGS STORAGE AND WATER MANAGEMENT					
Activity	**Deliverable**	**Description**	**SS**	**PFS**	**FS**
Governance	Tailings management policy, RACI charts and review reports	Policy required before the commencement of study to inform the design of the company's internal requirements. Independent review recommended at Scoping level but essential by Pre-Feasibility and into Feasibility.	✓	✓	✓
Tailings technology studies	Report, including maps and drawings	Assessment of technologies that can be used to dispose of tailings. The assessment should take into consideration the throughput, tailings physical characteristics, geochemistry, site topography, climate and water demands. Several technologies may be carried forward from Scoping to Pre-Feasibility, with the outcome of the Pre-Feasibility Study being selection of the preferred disposal technology.	✓	✓	
Siting studies	Report, including maps and drawings	Assessment of potential tailings disposal site and water storage options through multiple account analysis, which should consider social, environmental, technical, risk and financial considerations. Several sites/options may be carried forward from Scoping to Pre-Feasibility with the outcome of the Pre-Feasibility Study being selection of the preferred sites/options.	✓	✓	

TABLE 7.1 CONT...

Activity	Deliverable	Description	SS	PFS	FS
Risk assessment	Risk matrix, failure mode and effects analysis, risk report	At all phases, qualitative risk evaluations should be conducted and where justified by the risk profile of the facility, semi-quantitative or quantitative risk evaluation should be included by the Feasibility Study.	✓	✓	✓
Dam breach assessment	Report, including maps	Assessment of potential downstream impacts in the event of a dam failure. Complexity generally increases at each study phase and for projects where preliminary studies indicate a higher level of risk or where the study will feed into the development of emergency response plans or resettlement planning.	✓	✓	✓
Consequence Category assessment	Report	Consequence Category assessment should be conducted at each phase of study as the understanding of potential impacts in the event of a dam failure are advanced and as engineering designs are applied to reduce risk or eliminate potential failure modes. This is the key activity required to define the design parameters for the tailings storage and water dams.	✓	✓	✓
Hydrology	Analysis and report	The site's climate is one of the key drivers for the design and impacts the selection of suitable sites/technologies, embankment design, water balance, spillways and closure. At the Scoping phase, basic monthly rainfall data is generally all that is required along with identification of suitable data sources for future study phases. The installation of site weather stations to collect data if insufficient pre-existing data is available for the site should be a priority. As the design develops though the Pre-Feasibility phase then development of intensity frequency duration curves, estimates of probable maximum precipitation should be conducted and by Feasibility, either long-duration historical daily record or a synthetic stochastic data set should be generated.	✓	✓	✓
Geotechnical site investigations	Report, including logs, photos and laboratory test report	Increasing in complexity and detail as the study progresses. At the Scoping Study this is commonly limited to desk-top studies of geology and seismicity and visual mapping. At Pre-Feasibility it should include some physical investigation of proposed site (test pit and/or borehole). By Feasibility, it should include comprehensive field and laboratory programs to fully characterise the foundation and potential construction materials.	✓	✓	✓
Tailings characterisation	Test work and report	At the Scoping phase, is generally limited to review of metallurgical testing and estimation of geotechnical and geochemical characteristics based on historical case studies. At Pre-Feasibility, physical testing of the tailings should commence to determine their settling, drying and geochemical behaviour. Additional samples should be tested at Feasibility to define the variability of the material over the life-of-mine and should include both static and kinetic geochemical testing.		✓	✓
Facility design	Report, including maps and drawings	A conceptual selection of the preferred embankment type is to be made at the Scoping Study, with the viability of the selection to be tested at Pre-Feasibility as more data on the site conditions becomes available. Deposition planning, layouts, embankment type, staging, tailings delivery and water return systems to be optimised and finalised at Feasibility to fully account for the site conditions.	✓	✓	✓
Liners and underdrainage design	Report, including maps and drawings	The conceptual requirement for liners should be defined at the Scoping Study phase using conservative assumptions. At Pre-Feasibility this can be revised as more data on the tailings chemical characteristic and seepage potential is available. At Feasibility, the exact types and extent of the liners should be fully defined based on a well-defined knowledge base.	✓	✓	✓

TABLE 7.1 CONT...

Activity	Deliverable	Description	SS	PFS	FS
Seepage analysis	Analysis and report	At the Scoping phase, insufficient data is normally available to undertake meaningful assessments. By Pre-Feasibility, sufficient data should have been developed on the tailings and foundation to allow for 2D seepage analysis. By Feasibility, for complex geometries or high-risk facilities a preliminary 3D groundwater seepage model should be developed which will be expanded and refined in detailed design and throughout operations.		✓	✓
Stability analysis	Analysis and report	Similar to the seepage analysis, limited data is commonly available at the Scoping phase, but some high-level stability using database material parameters should be conducted. At Pre-Feasibility, limit equilibrium stability analysis at the critical embankment section should be undertaken. By Feasibility, multiple sections should be analysed using limit equilibrium techniques with finite element or finite difference modelling to be undertaken for geotechnically challenging or seismically active projects.		✓	✓
Dynamic deformation analysis	Analysis and report	Generally insufficient information available to undertake meaningful analysis at the Scoping phase. At Pre-Feasibility, empirical calculations can be undertaken as well as a site-specific seismic hazard analysis. For sites with a high seismic risk, complex finite element or finite difference modelling should be included by Feasibility.		✓	✓
Water balance	Analysis and report	The Scoping Study can be limited to high-level monthly water balance to determine approximate site demands or excesses. At Pre-Feasibility, a shorter time step model run either with deterministic rare events or probabilistically. Feasibility should have a model that considers all potential variables probabilistically.	✓	✓	✓
Surface water management	Report, including maps and drawings	At Scoping, is limited to schematic layouts with limited sizing. Pre-Feasibility should include peak flow estimation and hydraulic design. Feasibility should integrate the geotechnical design of the structures and include detailed hydraulic modelling.	✓	✓	
Operational, maintenance and surveillance (OMS) manual	Reports, including maps	From Pre-Feasibility, develop an OMS manual that ensures the facility functions in accordance with its design intent, meets performance objectives and regulatory and corporate policy obligations, supports the risk management plan and links to emergency preparedness. Finalise at Feasibility with input of operational personnel.		✓	✓
Instrumentation and monitoring	Report, including maps and drawings	At the Scoping phase, an allowance for cost should be made. By Pre-Feasibility, instruments should be installed to provide baseline data required for environmental permitting. At Feasibility, plans should be in place for a comprehensive monitoring network to assess the performance of the tailings storage facility (TSF) construction and operation against the performance objectives and indicators and the risk management plan. Trigger action and response plan (TARP) to be developed for instrumentation to address credible failure modes.	✓	✓	✓
Closure plan	Report, including maps and drawings	Required from project concept and should guide the design in each study phase. Increasing level of detail in each phase of study; however, it is unlikely that the TSF closure system design can progress past a Pre-Feasibility Study level of detail and accuracy (both in terms of material requirements and costs) prior to the facility becoming operational.	✓	✓	✓

SS = Scoping Study; PFS = Pre-Feasibility Study; FS = Feasibility Study; RACI = responsible, accountable, consulted, informed.

GOVERNANCE

Good tailings governance is a fundamental requirement of any tailings management system throughout the entire life cycle from conception to post-closure and includes all study phases. Effective governance provides an essential foundation for all activities related to tailings management. The goal of good governance is therefore to manage risk. In the same way, as part of that governance, independent review provides a 'clean pair of eyes' to provide checks and balances and to ensure that all areas of potential risk have been evaluated and understood, even if they are not possible to remove.

For mature mining companies a governance structure will be in place, and this will provide a template for the designer to undertake the studies. For members of the International Council on Mining and Metals (ICMM) at least, this governance framework will need to comply with the *Global Industry Standard on Tailings Management* (GISTM) published by the Global Tailings Review (GTR, 2020). With governance in place, a mining company will have a policy that will have been defined and agreed at a senior corporate level that provides the designer with the appropriate guidelines and standards to which they must adhere. Not only will this cover engineering guidelines such as those provided by the Australian National Committee on Large Dams (ANCOLD) and Canadian Dam Association (CDA), but also cover those areas such as social and environmental sustainability, human rights, community engagement, public health, etc. The policy should also set out a fundamental set of principles and standards that relate to the design and tailings management as laid out by various bodies such as the ICMM, GTR, CDA and Mining Association of Canada (MAC), together with key decision points (gates) throughout the design process.

It is important that policy, standards, principles and objectives are understood at each level of study, particularly from the outset. Good governance should always be adaptive, and the potential advances leading to changes that are accepted in the industry should be considered in any study report. Consequently, at every level of study the policy, standards, principles and objectives should be re-evaluated.

It is also important from the outset to understand the competencies required for tailings management so that the appropriate personnel (eg employees, consultants) have those competencies.

A fundamental component of the governance function of studies is review and assurance and particularly, independent review. The GISTM indicates explicitly, for instance, that the external independent review for technical matters should be established early in the life cycle and throughout all phases. The GISTM also states that for specific elements of the TSF location selection or dam hazard classification, the independent review shall be consulted.

The input of independent review should be sought from the initial conceptual planning and design phases, through to reviewing post-closure performance. However, over the life cycle of a tailings facility, and as the risk profile of a facility changes, the scope and focus of independent review should be re-adjusted to ensure it remains relevant and effective. It is also important that any independent review allows for future mine development and associated tailings management beyond current life-of-mine planning as adopted in studies. A difference of even a few years can have a significant impact and implications for tailings management and design.

GENERAL PRINCIPLES FOR TAILINGS STORAGE AND WATER MANAGEMENT

The overarching objectives of the tailings storage and water management design process are to design a system that minimises risk to persons and the environment, is financially attractive, technically viable and can be effectively and safety closed at the end of operations. To achieve these objectives, the following key principles should be adopted in the design:

- maximise the recycling and reuse of water and minimise the generation of excess mine-affected water
- develop a robust design that provides surety of water supply and is sufficiently sized to accommodate extreme rainfall and climatic conditions that may be experienced over the project life
- maximise tailings density unless there is a specific design requirement to maintain saturation of the tailings
- develop a system that is geotechnically stable under all loading conditions both during operations and closure
- minimise the environmental impacts of the facility whether by seepage, airborne emissions or overland discharges throughout the tailings life cycle
- develop a tailings storage management system that accounts for the geochemistry of the tailings and minimises the potential for adverse chemical reactions both during operation and at closure
- develop a design that considers long-term closure at its core, with the goal of developing long-term stable landforms at closure
- develop a monitoring system that allows the performance and safety of the system to be continually assessed and for early detection of any adverse conditions.

GUIDELINES AND STANDARDS

Tailings storage and water management systems are covered by local (state/territory), national and international guidelines and standards. In many cases the mining company will require that the design complies with recognised international guidelines such as those published in Australia by ANCOLD, in Canada by MAC or CDA, or internationally by ICMM, Global Tailings Review or the International Commission on Large Dams (ICOLD), irrespective of the jurisdiction where the project is located. In the absence or lack of regulations in a given jurisdiction, the designers should apply appropriate measures to international standards.

National standards in some countries or jurisdictions prohibit or limit the use of certain management practices that are commonly employed in other jurisdictions (such as upstream construction, subaqueous tailings closure or release of mine-affected water) and it is commonplace for conflict to exist between the various guidelines.

At the Scoping Study phase, it is imperative that all guidelines, standards and regulations that may apply to the project be reviewed and used to establish a design basis. Where a conflict is found between the guidelines then the more stringent standard should normally be adopted. However, additional design or analysis may need to be conducted during this phase to demonstrate that the design complies with all applicable standards.

The design basis should be documented in a design basis report (DBR), which will be a live document and is to be updated and revised over the course of the study phases as the design evolves, and the life of the project.

TAILINGS TECHNOLOGY STUDIES

Best available technologies (BATs) and best available practices (BAPs) represent the best available techniques to manage tailings and water for a specific project. The best overall approach for a project is typically a combination of BATs and BAPs that most effectively reduce the physical, geochemical, ecological, social and economic risks associated with the entire storage life cycle, from construction through operation and closure.

BATs are not a 'one size fits all' solution and are specific to project requirements, climate, geology, geomorphology, downstream sensitivity, tailings characteristics and social considerations. Site selection considerations, technologies and design features that provide for resilient and robust tailings storage and water management throughout a mine's life cycle are considered when defining the BAT for a project.

BAPs for the project are the management and operational procedures that are developed with consideration of engineering and governance practices for the most appropriate BATs. The BAPs should confirm that the tailings and water management system is designed, constructed, operated, maintained, monitored and closed to achieve the performance objectives.

There is a move away from the construction of conventional slurry tailings dams with more projects adopting disposal of filtered tailings and disposal of tailings below ground (in-pit or underground backfill) to reduce the risk of catastrophic failure. When undertaking a study into tailings and water management the following are possible technologies that may be included in the assessment:

- conventional slurry tailings
- thickened slurry tailings
- paste (ultra-thickened) tailings
- filtered tailings
- cycloned tailings
- tailings and waste rock co-disposal
- underground backfill
- in-pit tailings disposal.

Each of these technologies will impact the site water requirements as the amount of water that will be left entrained in the tailings or lost to evaporation will vary considerably with the various options. Therefore, the tailings and water management components should be assessed as single system. Further guidance on available technologies can be found in ICOLD (2022).

Assessment of the BATs is normally a phased approach that follows the flow chart provided in Figure 7.1. At the Scoping phase, all potential technologies are considered with a prescreening assessment to be undertaken to eliminate any technologies that are fatally flawed. The initial prescreening can be a qualitative assessment undertaken by a design team with sufficient experience in tailings and water management who are familiar with the constraints of the project. This will reduce the number of technologies to allow further assessment to be undertaken on those which show merit.

Final selection of the preferred option (or options) requires the use of a documented multiple account analysis. Good guidance on undertaking a multiple account analysis for selection of BATs can be found in *Guidelines for the Assessment of Alternatives for Mine Waste Disposal* published by Environment Canada (2013).

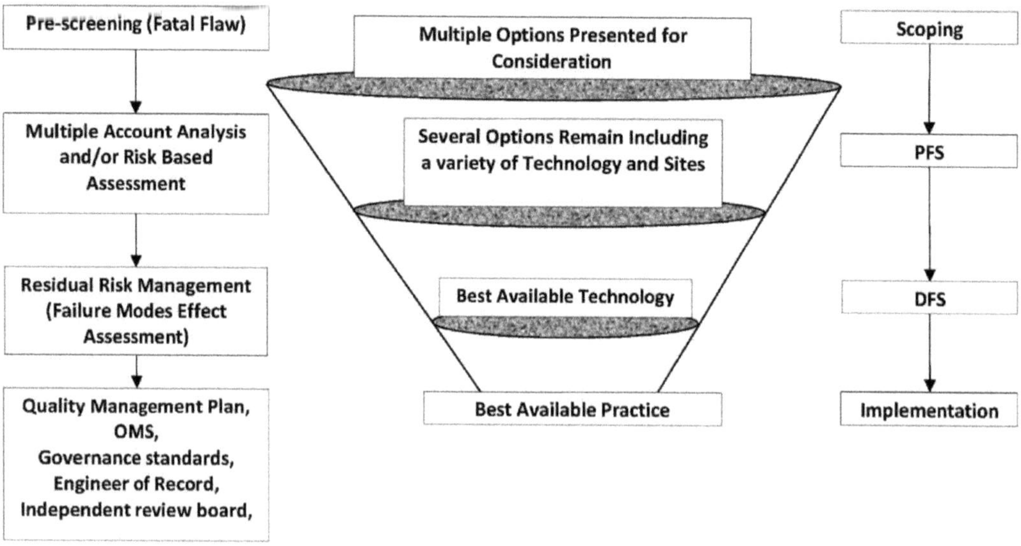

FIG 7.1 – Flow chart for the selection and implementation of best available technologies (BATs) and best available practices (BAPs).

SITING STUDIES

Siting studies should be undertaken at the Scoping Study phase to identify all sites that have the potential to store some or all of the life-of-mine tailings. It is important to consider the possible technologies that may be adopted as some sites may be suitable for one particular technology but inappropriate for another and some sites may have several potential uses (such as waste dumps, water dams or tailings disposal sites). This siting study should be undertaken in consultation with the process plant, infrastructure and mine design teams to ensure that competing demands are addressed. The site selection study should consider sites for tailings disposal, water storage, water harvesting and surface water management aspects. An overall site layout plan for the mine development for all potential options/technologies should be developed.

Generally, aspects such as environmental, social and land ownership / tenement constraints should be excluded from the initial development of potential site layout plans, so as not to eliminate options that show good technical merit and/or higher dam safety potential. It should be remembered that at Scoping Study phase it may still be possible for the mining company to renegotiate access to a particular site(s) that provides the best technical solution for long-term tailings and water management rather than accepting and concentrating their efforts on a technically inferior but 'convenient' option. It is beholden on the designer of a tailings and water management system to always remember that tailings (and the risks associated with them) will be there for the long term and should be planned accordingly.

For each option/layout, conceptual design should be undertaken to determine the disturbance areas and potential impacts, technical merits/challenges, high-level capital/operating costs, high-level closure concepts, high-level water balance and conceptual breach impact area (flow path assessment). Final selection of the preferred option (or options) requires the use of a documented multiple accounts analysis referenced previously.

Following the siting study, it is common to advance a single option to Scoping-level design in order to demonstrate that the project has at least one viable site for tailings and water management and the project merits advancing to the next phase of study. This option should not be considered to be fixed at this phase of the study and the designer needs to keep an open mind and be receptive to other options that may be proven in the Pre-Feasibility Study to be the preferred solution.

At the Pre-Feasibility Study phase, the design should reassess several preferred options in a greater level of detail when more data is available for the geology, geotechnical conditions, hydrogeological, social and environmental aspects. At the Pre-Feasibility Study phase of assessment, stakeholder engagement should have commenced such that project-affected communities are meaningfully engaged to provide input on the potential options. The multiple account analysis should then be re-run to select a single option that balances the competing technical, social, environmental and financial accounts to select a single preferred option to take through to Feasibility. By the commencement of the Feasibility Study the site layout is normally fixed, and no further site selection work is required.

RISK ASSESSMENT

The overall risk management framework for studies is covered in Chapter 11 but the risk management practices specific to tailings and water management facilities are summarised in this section for completeness.

The principles of leading practice tailings and water management are underpinned by a risk-informed decision-making applied to the planning, design, construction, operation, closure and rehabilitation of TSFs and water dams (ICMM, 2021). In this approach, plans need to be tailored to manage the TSF effectively over its full life cycle, with sufficient detail to manage the potential risks within acceptable limits.

Various risk analysis methods are applied to tailings and water facilities, which can be broadly split into qualitative risk evaluations, semi-quantitative and quantitative methods, and these are discussed comprehensively in ANCOLD (2022). Lying at the core of risk analysis for tailings and water management facilities is the identification of potential failure modes (hazards or threats), determination if they are credible failure modes and then the likelihood of occurrence if they are credible. Although a failure mode may be considered to be non-credible at an early stage of the design, it is important that this assessment be revisited at frequent intervals through the life cycle of a tailings facility to confirm the veracity of this assessment.

At the early stage of design (Scoping Study and Pre-Feasibility Study), risk analysis is generally limited to qualitative risk evaluations that are used to identify hazards, their likelihood, consequence, risk ranking and remedial actions. The remedial actions may be those already undertaken at that phase of study, or the analysis may identify those actions that are required in future phases of study (or implementation) to ensure that the identified hazards can be managed.

Semi-quantitative or quantitative methods of risk analysis may be undertaken at the Feasibility Study phase when the design has been advanced to a high level of study and a robust knowledge base for the site has been established. These methods can require large amounts of data to be collected to support complicated probabilistic interrelationships that may lead to the catastrophic failure of a dam. These semi-quantitative methods are required in some jurisdictions, such as New South Wales, to permit the dam. The aim of these assessments is to optimise the design to reduce the risk 'as low as reasonably practicable' (ALARP) and certainly to within the minimum acceptable safety threshold. Details of the requirement for these risk assessments can be found in societal and individual risk rating methodology for Dam Safety New South Wales (2022) and ANCOLD (2022).

DAM BREACH ASSESSMENT

Dam breach assessment and the potential impacts of a failure are the key input to the Consequence Category assessment and hence the design criteria that will be applied to the tailings or water dam designs. This is why dam breach assessments should be undertaken, albeit at a high level, from the Scoping Study phase.

At the Scoping Study phase, when various options are still being assessed, flow path analysis should be undertaken to define the direction of flow in the event of a breach. Conservative assumptions on the potential extent of the inundation along the flow path should be made in order to allow the potential impacts on infrastructure, communities, the environment and the population at risk to be estimated.

Moving to Pre-Feasibility, each option under assessment will require a dam breach assessment to be undertaken; however, the complexity of the model can be limited unless the level of uncertainty of flow path analysis is high. If, for example, it is clear from the simplistic flow path assessment what order of magnitude the population at risk (PAR) will be, there is no need to complete more detailed modelling at these early phases. Modelling should include analysis of credible failure modes and a run-out analysis. At the Pre-Feasibility Study phase, conservative assumptions usually need to be made regarding breach development, tailings liquefaction, tailings rheology and the contribution of free water.

As the project moves into the Feasibility Study (and continuing into the implementation phases of the project), the dam breach assessment provides information to assist in developing an emergency response plan. This requires not only the area of impact to be defined but also the likely time to inundation, flow depth and flow velocity so that evacuation plans, warning and alarm systems can be designed. This requires complex modelling of the breach outflow using 2D dynamic modelling, which may also include non-Newtonian flow to simulate the release of liquefied tailings. It is noted that while breach formation can be determined, historical failures such as the Brumadinho dam failure in Brazil in 2019, have shown that liquefaction can occur rapidly with little to no warning and the resulting hydrograph is unlike that seen with typical overtopping or piping failures. The possibility of such an event occurring at the facility should be considered when determining what breach hydrograph should be adopted.

A comprehensive *Technical Bulletin on Tailings Dam Breach Analysis* has been released by the Canadian Dam Association (2021). It is considered the current leading practice for undertaking dam breach analysis.

CONSEQUENCE CATEGORY

A Consequence Category is assigned to a dam or facility according to the population at risk or potential loss of life, the magnitude of the adverse consequences affecting the community/ business interests and environmental effects, which could be expected to result from that facility's failure or release from a facility. The Consequence Category of a dam is independent of the likelihood of a dam failure and can change over time with changes to downstream populations, infrastructure or environmental values. The Consequence Category can also change over the life of the project (or design phase) as the dam is modified (eg raised) and therefore will need to be updated at each phase of study where there is a material change to the tonnage of tailings or volume of water to be stored in the facility.

As the Consequence Category defines many of the design criteria used in the design of a tailings or water management facility, it should be undertaken at the Scoping phase and updated through each phase of design. However, current best practice (GTR, 2020) requires that the ability to upgrade a tailings storage facility to comply with an Extreme Consequence Category classification is demonstrated to concept level throughout the TSF life cycle. In this context, the precautionary principle should be adopted, and design parameters commensurate with an Extreme Consequence Category classification should be adopted until sufficient design detail is available to demonstrate otherwise.

Various methods are available for determining Consequence Category including those of ICOLD (2022), GISTM (GTR, 2020), ANCOLD (2012, 2019a) and individual states such as Queensland (DES, 2016). As there are sometimes variations in the design criteria defined by each method it is recommended that the most conservative design criteria are adopted for the various methods used to determine Consequence Category.

In addition to the Dam Failure Consequence Category, which is used to determine the design parameters that affect the stability and integrity of a tailings facility such as the design of earthquake loads and spillway capacity, both ANCOLD (2019a) and DES (2016) include additional consequence categories. These are the ANCOLD Environmental Spill Consequence Category and DES Failure to Contain – Overtopping Consequence Category, which relate to the containment of mine-affected water within a tailings or mine water storage dam, and the DES Failure to Contain – Seepage Consequence Category relating to seepage from tailings or mine water dams.

The author recommends that engineering judgement should be applied in selection of design parameters based on the average recurrence interval for very long-life projects. Although it may be acceptable to employ a 1 in 100 year average recurrence interval (ARI) spillway release for a facility that will have an operating life of a decade (which would result in the probably of release over the life of the project of approximately 9.5 per cent), applying the same ARI with a facility with an operating life of 50 years would increase the probability of release to approximately 40 per cent over the life of the project, which may be unacceptable to the mining company and other stakeholders.

HYDROLOGY

The analysis of rainfall data relies on the sourcing of suitable data, which can consist of monthly, daily and subdaily interval data sets. Monthly data sets are generally the easiest to source, especially in developing countries. Daily data sets are not as easy to source but generally can be acquired; however, subdaily data sets are often extremely difficult to find, especially in developing countries. The nearest rain gauge to the project site may be some distance away and may not be representative of the site conditions. As such, it is recommended that a weather station, capable of taking readings (both rainfall and wind speed/direction) at a minimum 15-minute interval, be installed as soon as is practicable at the project site. The number of rain gauges and their location to be installed should take into account the overall extent of the project area, as well as the possibility of rainfall variability over the site due to phenomena such as the orographic effect.

Data sets that are sourced should be checked to ensure the quality and the completeness of the data set is acceptable. The largest rainfall events on record should be cross-checked against news reports and other publications to confirm their validity. Nearby stations should also be checked to see whether the same significant rainfall event occurred at them as well.

Rainfall analysis, for the purpose of tailings dam and water management design, can be broken up into two parts, the determination of long-duration rainfall events and the determination of short-duration rainfall events. For long-duration events, the outputs are typically average annual rainfall, average pan evaporation and then statistical analysis to determine the probability of experiencing extremely wet or extremely dry years. For this part of the analysis monthly data sets can be acceptable. For the short-duration extreme rainfall events, daily precipitation data can be used to determine the nth day design rainfall depths.

Scoping designs will normally be undertaken based on monthly data sets or, if available, daily data sets. As the design progresses a more detailed data set should be developed including subdaily data as well as generation of stochastic daily rainfall data sets. For projects located in regions without specific methods for developing of intensity-frequency-duration curves, subdaily precipitation data is critical for determining design rainfall depths with durations less than 24 hours. Subdaily data is also required for providing guidance on the selection of representative storm hyetographs. Where subdaily data is not available, preliminary estimates of subdaily design storm depths can be made using available methods that estimate the subdaily storm depths from the 24 hour rainfall depths. However the installation of a suitable weather station at the project site should be prioritised.

The potential impact of climate change on extreme rainfall events should be considered when assessing rainfall data. The potential impact of climate change on long duration rainfall, including average annual rainfall, and evaporation should also be considered, particularly in relation to closure design, and its long-term performance. Rainfall data sets should be reviewed at a regular basis during the tailings facility life cycle to identify potential changes and update design criteria accordingly.

GEOTECHNICAL SITE INVESTIGATIONS

Geotechnical site investigations are undertaken to determine the characteristics of the foundations and potential construction materials for the dams, channels, spillways and other components of the tailings and water management system. They are required for all new dams no matter the size of the dam.

Geotechnical investigations may require the input of engineering geologists, geotechnical engineers, rock mechanics engineers, hydrogeologists and civil engineers to ensure the investigations are targeted to the type of facilities that are intended to be constructed and to target information that can help address the potential failure modes that have been identified in the risk assessments. Comprehensive guidance on the geotechnical investigation for dams can be found in ANCOLD (2020) or APEGBC (2016).

Over the study stages the investigations are normally phased, as shown in Figure 7.2, with the investigation at Scoping normally limited to desk-top and literature reviews to identify the likely geological regime and soil types expected, followed ideally by walkover inspections of the proposed sites by an experienced engineer or engineering geologist to identify fatal flaws or obvious adverse conditions.

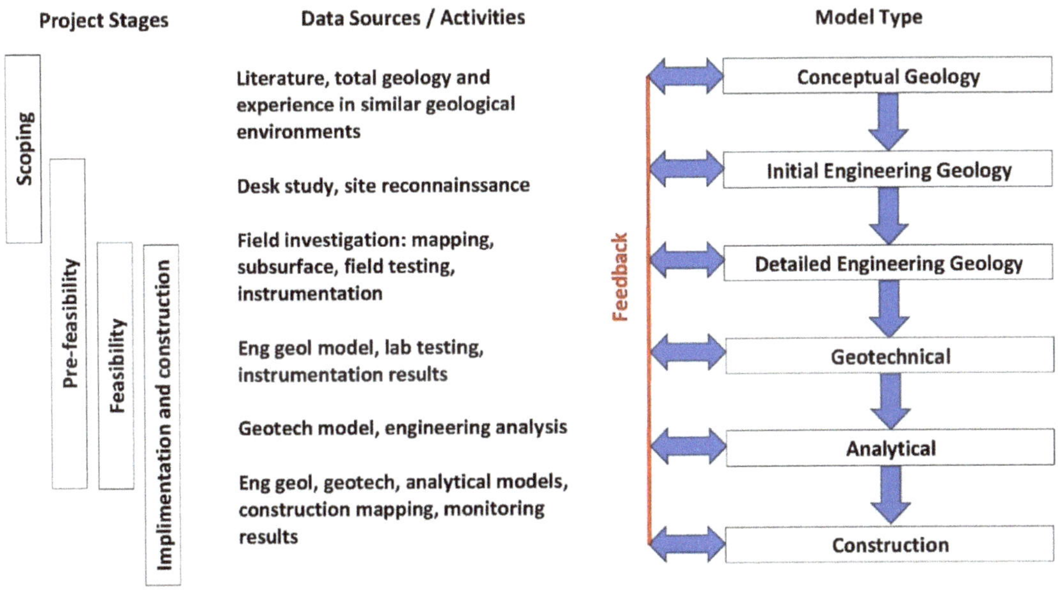

FIG 7.2 – Model types and project phases (adapted from Figure 2 in the ANCOLD Guidelines for Geotechnical Investigations of Dams, their Foundations and Appurtenant Structures. ANCOLD, 2020).

At the Pre-Feasibility phase, the field component should increase to include physical investigation such as test pits and/or boreholes, field mapping and installation of instrumentation to collect data on groundwater conditions, which will be required at the start of the Feasibility Study. Laboratory testing at the Pre-Feasibility phase is normally limited but may include basic geotechnical testing such as index properties, permeability and compaction testing for potential construction materials. For large projects or if problematic foundations are suspected, then a limited amount of more complex geotechnical testing such as triaxial, oedometer or direct shear may be warranted to confirm the viability of the proposed site. Geophysical investigations along the full length of the proposed embankment or channels at Pre-Feasibility to link and correlate between a limited number of wide spaced boreholes can be a cost-effective, low-impact and expedient method to improve the confidence in the data provided.

At Feasibility, a more focused and comprehensive geotechnical investigation program should be implemented which will target specific structures for the dams and water management components and construction materials sources. At Feasibility, both *in situ* testing (permeability and strength) and laboratory testing to calibrate the *in situ* testing should be undertaken. A database of geotechnical properties, and their variability, for all the anticipated materials types, which are either present within the foundation or are anticipated to form the construction materials, should be developed. The actual testing to be undertaken will be site specific and should be intended to address the potential failure modes identified for the design and should include assessment of shear strengths, soil behaviour types (drained or undrained, brittle/contractive or dilative), liquefaction potential, permeability characteristics, compressibility, dispersion and piping potential, collapse potential and in high-seismic regions, dynamic performance under earthquakes loads.

At each phase of investigation, a factual and interpretative report for the site(s) should be prepared and this should be updated with each phase of additional investigation so that a single compendium of all geotechnical data is maintained for the site(s) for future reference.

TAILINGS CHARACTERISATION

Tailings characterisation involves the determination of both the physical and geochemical properties of the tailings. Tailings are produced by a multitude of process flow paths from ore material derived from all types of rocks. They therefore have a large degree of variability that needs to be considered and understood during each study phase. The tailings characteristics will be one of the key inputs into selecting the best available technology and water management strategy for the project.

When a Scoping Study commences only limited metallurgical testing will have been undertaken, generally only enough to define a potential process path and preliminary indications of the likely recovery achievable from the ore. The metallurgical testing will be ongoing throughout the study phase, optimising the processing path and factors such as grind size and process reagents. Even the process method can change as the study progresses. Therefore, the tailings and water management design will need to be continually updated to reflect these changes.

At Scoping Study, it is common that a very limited amount of tailings material will be available for testing, predominantly due to the scale of the test work undertaken at this phase of study. At Scoping phase, it is therefore normal for the physical and geochemical characteristics of the tailings to be assumed or estimated from basic soil testing rather than being physically tested. Tailings physical parameters suitable for use at Scoping phase can be sourced from various literature sources such as McLeod and Bjelkevik (2017) and ICOLD (2022).

At the Pre-Feasibility phase, small metallurgical samples of tailings are normally available, which allows for laboratory testing to commence. The laboratory testing program for the Pre-Feasibility phase should include index properties (gradations and Atterberg limits) and undrained settling testing as well as acid–base accounting, whole rock multi-element and supernatant chemistry testing. It is important at the Pre-Feasibility phase to ensure that planning is in place to produce sufficient samples for the Feasibility Study, with the test work program from Feasibility requiring a minimum of 20 to 40 kg of each tailings type. This means that hundreds of kilograms of ore material will need to be processed through bench-scale tests to generate the required quantity.

At the Feasibility Study phase, multiple samples of tailings either from time-period metallurgical composites or ore-type metallurgical composites should be tested. The testing at Feasibility should be comprehensive and tailored to the tailings technology and disposal method being adopted. Examples of the test work typical of a Feasibility Study are provided in Table 7.2.

The results of this testing should be interpreted to provide the tailings parameters for use in the deposition modelling (beach slopes), water balance modelling (water release), stability (shear strength), capacity assessment (tailings density), geochemical models (leach rates), tailings pumping and dewatering systems and closure design (capping/cover requirements).

TABLE 7.2

Feasibility Study comprehensive tailings laboratory testing program.

TAILINGS DEPOSITION AND SETTLING	TAILINGS GEOTECHNICAL PERFORMANCE	TAILINGS GEOCHEMICAL CHARACTERISTICS	TAILINGS PUMPING AND DEWATERING
• Index properties • Particle density (SG) • Undrained settling testing • Drained settling testing • Air drying • Permeability and consolidation parameters	• Strength testing (triaxial, direct shear, direct simple shear, critical state triaxial, cyclic strength testing) • Compaction testing (proctor)	• Acid–base accounting and net acid generation • Whole rock and supernatant multi-element analysis • XRD • Leach extraction testing (shake flask or TCLP) • Kinetic testing (column leach or humidity cells)	• Tailings thickener testing • Tailings filtering • Tailings rheology

SG = specific gravity; XRD = X-ray diffraction; TCLP = toxicity characteristic leaching procedure.

FACILITY DESIGN

Having established the design basis (documented in the DBR) and identified potential site(s) for one or more disposal technologies during the Scoping Study, the following activities should be undertaken as part of the Pre-Feasibility Study design process and evolve through the design process as more data becomes available to be finalised by the completion of the Feasibility Study:

- deposition or emplacement planning
- optimisation of layouts
- embankment type (in the case of above ground flowable waste)
- staging of development
- tailings distribution method
- water return (if required).

The selection of the embankment type (upstream, centreline and downstream) should commence at the Scoping Study phase as the volume of construction material required, hence the capital cost for tailings storage, is highly dependent on the embankment type. However, the selected embankment type should be reaffirmed at each phase of study as more site and tailings characterisation data become available, as this information will either reinforce the selection or may invalidate the decision and require redesign.

Figure 7.3 provides an overview of the strengths and weaknesses of each of the embankment types and illustrates that selection of an embankment type is site specific and requires consideration of the site setting as well as the tailings characteristics. Historically, upstream tailings dams have had higher failure rates than other types of embankments, and although there have been significant improvements in understanding the geotechnical behaviours of tailings and appropriate construction techniques over recent years, the consequence of a failure should also be considered during the selection of the embankment types.

Where the TSF may be constructed in stages, the Scoping Study should focus on the final arrangement of the facility. From the Pre-Feasibility Study, both the initial and final arrangements should be developed to a level commensurate with the phase. During the Feasibility Study, intermediate stages need only be developed to a level commensurate with a Pre-Feasibility Study, while plans and analysis of the closure arrangement are typically only developed to a concept level.

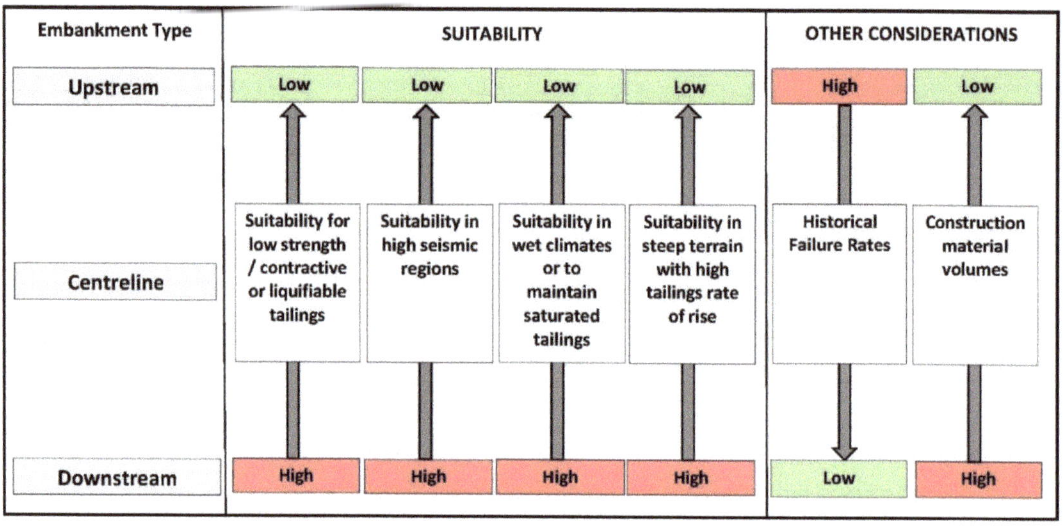

FIG 7.3 – Considerations in selection of embankment type.

The embankment design for facilities that are designed to store water, either those tailings dams that are intentionally designed for this purpose or water storage facilities, should be designed to follow the geotechnical principles and dam technologies developed from the water dam community.

LINERS AND UNDERDRAINAGE DESIGN

Seepage from tailings and water storages can lead to groundwater contamination, undesired changes to groundwater levels and flow regimes and to instability in the confining embankments. Good design of seepage mitigation measures during the study phase reduces the likelihood of retrospective seepage reduction measures such as grout curtains, seepage interception trenches or recovery bores being required during operations. These retrospective measures are commonly less efficient, costly and can require ongoing post-closure active management.

In some circumstances seepage from tailings storages and water storage facilities does not need to be controlled where excess water is available, the water is benign and modification to the groundwater regime would not impact stability or downstream receptors. The decision on what level of seepage control is required should be risk assessed based on the geochemistry of the tailings, chemistry of the mine-affected waters, sensitivity of the receiving environment, facility stability and availability of water. Tailings disposal techniques that minimise the amount of water entrained in the tailings (such as filtered tailings) will reduce the seepage potential but some seepage may still occur due to infiltration of meteoric water and water release on consolidation. This should be accounted for in the design.

Liners act to reduce the permeability of the storage facility basin, thereby reducing the seepage losses from the facilities. Liners can be broadly divided into synthetic lines and soil liners (including amended soils). Consideration of the longevity and chemical compatibility of the selected liner systems is required as saline, caustic or acidic conditions can alter the permeability and characteristics of some synthetic and soil liners.

Underdrainage systems have the effect of reducing the hydraulic head acting on the liner(s) and can be installed within the tailings or, in certain situations where double liner systems are installed, can be installed between liners to reduce the hydraulic head acting on the lower (secondary) liner. Where liners are included in the design and there is the chance of groundwater daylighting below the liner (ie springs being present), then subliner drainage will be required to relieve this water pressure below the liner.

At the Scoping Study phase, limited or no data will be available on the groundwater conditions at the proposed sites and data on the tailings geochemistry and process/mine-affected water chemistry may be limited. It is prudent in these scenarios to adopt conservative assumptions on the requirement to install liners and drainage systems and the extent of such systems. The ability to construct the liner should be considered during the site selection process.

At the Pre-Feasibility Study phase, the inclusion of liners and drainage should be reassessed based on the updated knowledge base for the sites and the geochemistry. By the Pre-Feasibility phase, modification to the process path may have been undertaken (such as inclusion of cyanide detoxification or removal of sulfides from the tailings) and these can negate the need for liner systems or reduce the extent or robustness of the required liner systems.

At the Feasibility Study phase, a full understanding of the site hydrogeology, tailings geochemistry and process/mine-affected water chemistry should be developed and the liner systems should be fully designed. This should include physical testing of the performance of the proposed liner materials with samples of water to be stored in the facility to ensure the liner material will result in acceptable performance. The drainage system should be designed to ensure filter compatibility between fine base soils and granular drainage. Where geofabrics are proposed as part of the design, the potential for clogging or degradation of the geofabric should be assessed at this phase of the design.

SEEPAGE ANALYSIS

Seepage assessments are generally conducted to quantify potential seepage through the tailings storage basin, reservoir rim, embankment and foundation, which will inform the seepage mitigation design (environmental compliance designs), and also be utilised to define the location of the phreatic surface and pore water pressure distributions within the embankment and foundations to support the stability analysis. Quantification of seepage losses from the basin and through the embankment will also inform the sitewide water balance modelling.

During the Scoping phase, no or very limited information related to site hydrogeology and tailings characteristics will be available and therefore it will be challenging to undertake a meaningful seepage assessment without reliable input parameters.

During the Pre-Feasibility phase, some information will be available from visual observations from the site visits, preliminary geotechnical and hydrogeological investigations and information acquired during installation of instrumentation for initial environmental monitoring. *In situ* and laboratory permeability testing on different soil lithologies encountered and packer testing on each rock weathering grade will be undertaken to define the hydraulic conductivity parameters utilised in a seepage assessment. At least one critical section of the embankments and saddle dams needs to be analysed at the Pre-Feasibility Study phase.

During the Feasibility Study phase, more rigorous seepage analysis, including 3D analysis, may be required to establish dam safety performance and environmental conformance. Additional site-specific hydrogeological data will be available during this phase to better define input parameters used during the Pre-Feasibility Study phase.

As part of the Feasibility Study, development of a detailed 3D groundwater model may be required to quantify the seepage through the TSF basin, perimeter ridgelines, embankment and saddle dams and also prediction of the flow paths (particle tracking study) to inform operational and closure sitewide water balance, contaminant transport and water quality modelling.

STABILITY ANALYSIS

During the Scoping phase, no or very limited geotechnical information will be available and therefore it will be challenging to undertake a meaningful stability analysis. Therefore, at this phase, the embankment batter slopes will be typically defined by designers' experience based on the likely material available for construction and anticipated foundation conditions.

During the Pre-Feasibility phase, some information will be available from visual observations from site visits, preliminary geotechnical investigations and information acquired during drilling for installation of instrumentation such as monitoring bores. *In situ* and basic laboratory testing on different soil and rock lithologies can be used to define drained and undrained shear strength parameters for 2D limit equilibrium stability analyses. Factors of safety (FoS) under short-term (undrained), long-term (drained) and post-seismic loading conditions are required. Rapid drawdown analysis of upstream embankment slopes is also required for water retention dams.

During the Pre-Feasibility phase, screening tests based on Atterberg limits and grading tests should be used to assess the potential for static or dynamic liquefaction in tailings where embankments are constructed by upstream methods and/or foundations consisting of contractive material. Where there is a potential for liquefaction, stability analyses should be undertaken using post-liquefaction strength parameters. Note that the pseudo-static analysis as a screening tool for earthquake stability is now not recommended by the best practice guidelines (ICOLD, in press; ANCOLD, 2019a).

Engineering judgement is required in selecting the methods of analyses and search routines offered by numerous software programs. At least one critical section (highest/steepest embankment section, area with the weakest foundation, section with the most severe consequences) of the embankments and saddle dams needs to be analysed at the Pre-Feasibility Study phase.

During the Feasibility Study phase, more rigorous stability assessment is required. Additional field and advanced laboratory testing data (possibly including critical state and cyclic strength testing) will be available during this phase to confirm the input parameters defined during the Pre-Feasibility phase. Engineering judgement must be used to include or exclude the inferior test results in the interpretation adopting appropriate confidence intervals. Should the ground topography vary significantly along the crest and upstream to downstream toe of the embankment, 3D stability modelling may be required as 2D analysis is mostly suited for plain strain problems.

In addition to the analysis methods used during the Pre-Feasibility Study (short-term, long-term and post-seismic), post-peak analysis should be conducted during the Feasibility

Study phase utilising residual strengths. Depending on the complexity of the embankment geometry, numerical analysis may be required to establish the behaviour of embankment for various stress paths, which are not possible to evaluate by the limit equilibrium analyses.

Where high variability and uncertainty exists in foundation conditions, construction materials and tailings characteristics, sensitivity analyses and/or probabilistic analysis should be undertaken. Probabilistic analysis includes determination of the probability distribution of FoS based on probabilistic inputs for material parameters that represent spatial variation and uncertainty associated with the material properties. The work of Silva-Tulla, Lambe and Marr (2008) is a valuable resource to understand the dependency of FoS and probability of failure on the quality of investigation, design and construction.

DYNAMIC DEFORMATION ANALYSIS

Dynamic deformation analysis is a critical component of the design of TSFs in areas of high seismicity or for those facilities that contain materials suspectable to liquefaction under earthquake loading. During the Scoping phase, no or very limited geotechnical information will be available to define the dynamic parameters therefore no dynamic deformation analyses will be undertaken. At this phase, an appropriate allowance will be made for the crest settlement based on designers' experience.

During the Pre-Feasibility phase, the crest deformation of the embankment following earthquake shaking may be estimated based on simplified deformation analyses such as Swaisgood (2003) and Pells and Fell (2003). The input parameters for these high-level deformation assessments can be obtained from regional probabilistic or site-specific seismic hazard assessment, which will provide a relationship between peak ground acceleration and return period for the site. These methods are not appropriate where liquefaction of embankment or foundation materials is anticipated to result in deformation.

Notwithstanding the perceived seismicity of the project area, a site-specific seismic hazard assessment should be undertaken during the Feasibility Study to define seismic ground motion parameters (including peak ground acceleration, earthquake magnitude and response spectrum, the proximity of active faults and the maximum credible earthquake). These parameters are typically determined using the results of probabilistic and deterministic seismic hazard analyses conducted by a seismologist, or appropriately qualified and experienced individuals.

During the Feasibility Study phase, detailed 2D or 3D dynamic deformation analysis may be required to be developed using either finite element or finite difference software to simulate the development of stresses, strains and pore pressures within embankment, foundation or tailings, which may provide key information such as lateral shearing of filters and acceptability of adopted freeboard, or liquefaction initiation. This analysis will typically use parameters gained through advanced laboratory testing and shear wave measurements obtained by field testing.

Dynamic deformation analysis is a specialist area and should only be undertaken by persons experienced in the field. Refer to ANCOLD *Guidelines for Design of Dams and Appurtenant Structures for Earthquake* (ANCOLD, 2019b) for further information.

WATER BALANCE

The water balance is the link between tailings management and water management for the site and is used to calculate the water demands, water excesses (releases) and volume of water stored within the TSF and other storages on the site under a range of climate conditions. The water balance model should not be considered a design output but rather a tool that should be used throughout the design. The water balance should be used to assess the impacts of design modifications on the overall performance of the system or to direct changes that need to be made to the design to meet the design objectives (such as inclusion of water diversions to optimise the system).

At the Scoping Study phase, the water balance model can be a simple monthly balance of expected flows into and out of the various storages on the site and can employ reference values for parameters such as run-off coefficients, basal permeabilities, supernatant release and process water demands. Scoping level water balances are predominantly limited to simple Microsoft Excel models and may consider average conditions as well as some deterministic water balance runs that considered extreme dry and wet conditions.

As more data become available at the Pre-Feasibility phase, including better definition of tailings parameters, geotechnical conditions, catchment characteristics, rainfall and evaporation rates, availability of water supplies and process water demands, then the water balance model should become more complex to provide a high level of certainty for the design. The Pre-Feasibility Study model may still be limited to a deterministic model, but preferentially probabilistic modelling should be undertaken at the Pre-Feasibility Study.

The water balance model developed at the Feasibility Study should include updates to the input parameters and should consider both the behaviour of the tailings as well as the overall site water balance. The tailings behaviours, such as achieved density and beach slopes, are not fixed and will change seasonally and between dry and wet years with changes in rainfall, evaporation and as the volume of water (area) in the supernatant pond in the TSF changes. This can affect the available stormwater capacity in the facility so should be considered at this phase of study. Sufficient data should have been collected by the Feasibility Study to allow natural catchments to be modelled using catchment run-off models such as the Australian Water Balance Model (Boughton, 2004).

The water balance at Feasibility should be run as a probabilistic model and should include variability in all factors that can impact the water balance such as climate, plant throughput, tailings composition, tailings per cent solids and raw water demands (which do commonly deviate from the design values during operation). Modelling the full range of potential variability provides a more robust water balance model, decreasing the probability that water shortfalls or water excess will occur. Modelling at Feasibility is normally run as a Monte Carlo simulation but caution should be exercised by the designer as some jurisdictions (such as Queensland, Australia) do have specific modelling methods that need to be employed.

At the Feasibility Study phase there is the potential for the water balance model to be extended to include water chemistry modules into the water balance to provide estimates of the water chemistry in the various storages and flows around the site. This would require the input of a geochemistry specialist to integrate the water balance model with chemical modelling software such as PHREEQC®, which simulates chemical reactions and transport processes in the water system.

It is also becoming commonplace to include modelling of climate change scenarios at the Feasibility Study phase to assess the impact of changing climate on the site's water balance. This is especially relevant for long-life projects or where the water balance model is run to assess post-closure conditions.

SURFACE WATER MANAGEMENT

Surface water management includes the diversion of water to or from the site or around TSF, provision of facilities to store water to meet the mine's water demands, provision of facilities to capture mine-affected water, provision of facilities for treatment (and potentially release) of sediment-impacted water, together with spillways to allow the safe release of excess water from tailings or water management facilities.

The requirements for each site will be dictated by the jurisdiction in which the project is located as some jurisdictions prohibit the release of mine-affected waters whereas others merely apply conditions for release based on water quality and quantity. In more water-scarce environments there may be restrictions on the amount of water that can be captured by the surface water management systems in order to maintain environmental flows and water availability for other water users. Only in rare circumstances, where facilities are designed as no-spill facilities and where extreme flood storage is provided such that overtopping is inconceivable, should a facility be designed without an appropriate spillway.

The capacity of spillways for both tailings and water dams are based on the failure Consequence Category with recommended spillway capacity provided in ANCOLD guidelines (ANCOLD, 2019a and 2000). Following consideration of business risks, responsibilities to the community and the operating life of the facility the owner/designer may decide to adopt a larger flood capacity (ie a higher safety standard) than the minimum required based on the assigned Consequence Category. In some instances, the cost of increasing the capacity of the spillway may not be substantial but can help to reduce the risk of an overtopping failure of the facilities and complies with the ALARP risk principle, which should be applied to tailings and water storage facilities.

At the Scoping Study phase, the surface water management system design will be at a high level and may include schematic layouts of the system element such as diversion channels, dams and weirs to either remove or collect water. High-level estimates of the efficiency of the system will be undertaken as part of the water balance, with design generally limited to the main cost items and scaled costs. Hydraulic calculations at Scoping would normally be limited to empirical equations to provide preliminary peak flows.

At the Pre-Feasibility Study phase, the design effort will be increased with better definition of short-duration events allowing more accurate estimation of peak flow using flood routing models such as HydroCAD® or HEC-HMS®. This will allow for sizing of the structures and preliminary design of revetment requirements. However, due to the lack of geotechnical data at this phase of study, excavation slopes and fill batter slopes and estimation of the proportion of hard rock excavation (if present) will still be preliminary and should be based on the designer's experience.

At Feasibility, a geotechnical investigation should be completed for the main surface water management diversion channels, allowing verification of the base soils and their dispersion potential to confirm revetment design and geotechnical parameters that inform excavation and slope design. At Feasibility, the design should include for culverts, crossings, inlet structures and outlet structures (spillways and stilling basins), with a complete 3D model of the structures to be developed. Depending on the complexity of the system, computational fluid dynamic modelling may be required at Feasibility.

It is noted that if the tailings or water storage facility has a diversion system upstream of it, for spillway sizing, it must be assumed that this diversion system has failed, unless the diversion system is designed to convey the same event or rarer than what the spillway is being designed to convey.

OPERATION MAINTENANCE AND SURVEILLANCE

An operation maintenance and surveillance (OMS) manual is intended to ensure that the facility functions in accordance with its design intent, meets performance objectives and regulatory and corporate policy obligations, supports the risk management plan and links to emergency preparedness (MAC, 2019).

Although deposition planning should commence at the Scoping Study phase and be further refined through the project phases, development of a draft OMS manual is usually undertaken during the Pre-Feasibility Study and finalised with input from operations personnel during the Feasibility Study prior to commissioning.

The OMS manual should include all operational controls that can influence the performance and risk management of the facility including tailings transport and placement, containment, water management and reclaim, erosion and dust control (Morrison, 2022), with comprehensive details for the development of such a manual provided in MAC (2021b).

INSTRUMENTATION AND MONITORING PLANS

As part of the permit application and environmental and social impact assessment processes, agencies generally require mine operators to evaluate the baseline conditions around mine sites including proposed TSFs. These are discussed in Chapter 2. In this regard, instrumentation and monitoring plans need to be developed during the Scoping Study phase to allow adequate time for installation and collection of data for multiple years prior to the commencement of the Pre-Feasibility Study. A cost allowance for the preliminary instrumentation will generally be made based on previous experience.

In addition to instrumentation required to evaluate baseline conditions, instrumentation is required to monitor the response of the tailings/water facility and foundations to both construction and operation as well as that required for reporting or regulatory compliance.

The ICMM (2021) *Good Practice Guide* notes that monitoring provides information on parameters and characteristics that cannot be detected through site observation or inspections, cannot be observed with sufficient precision or accuracy, or need to be monitored at a high frequency or continuously, with the objective of collecting data to assess the performance of the TSF against the performance objectives and indicators and the risk management plan. Instrumentation is at the heart of both the observational and performance-based approaches to design, construction, operation and closure. However, it should be noted that current instrumentation is not effective in providing sufficient warning against the onset of brittle failure, and in particular liquefaction events.

A discussion of the instrumentation available to monitor tailings and water management facilities is beyond the scope of this chapter, however Morrison (2022) places the purpose of instrumentation into three categories:

1. Critical instruments, addressing a particular failure mode and functioning as a primary trigger for a monitoring alarm.
2. Non-critical instruments, to provide a more detailed picture should any one of the critical instruments reach a trigger.
3. Compliance instruments, required for reporting and/or regulatory compliance. These can be either critical or non-critical instruments.

Although an allowance should be made for instrumentation and monitoring at the Pre-Feasibility Study, instrumentation and monitoring should be clearly defined at the Feasibility Study phase and encapsulated with trigger action and response plans (TARPS) in a preliminary OMS manual.

CLOSURE PLAN

Closure plans should be integral to the design of the surface water and tailings management systems. Closure of tailings dams and water infrastructure is challenging and can be one of the most difficult aspects of overall mine site closure. Designing for closure will influence site and technology selection and design alternatives considered at both Scoping and Pre-Feasibility Study phases and can often be the critical differentiator between potential options.

It is common during the study phase for the overall mine closure design to be undertaken as a separate study to the engineering design. However, it is preferable for the designer of the tailings and water management system to be integrally involved in the closure planning.

The solutions for establishing a safe and stable facility and/or landform include physical (geotechnical stability, erosion and settlement), chemical, environmental and social stability. All these aspects should be addressed in the early design when site location, tailings management structure and the deposition method is decided.

The closure concepts developed during the site and technology selection undertaken during the Scoping and Pre-Feasibility Studies should be developed through the Pre-Feasibility and Feasibility phases. However, it is unlikely that the TSF closure system design can progress past a Pre-Feasibility level of detail and accuracy (both in terms of material requirements and costs) prior to the facility becoming operational.

CONCLUSIONS

Tailings storage and water management are interlinked and are essential components of most mining projects. However, if managed poorly, they can lead to catastrophic failures, loss of life, irrecoverable environmental impacts and social dislocation. The design requires a team of specialist who have both the technical experience and specialised knowledge to undertake the study work. The tailings and water management design should always be undertaken with the view of minimising the risks ALARP and provide a clear pathway to sustainable closure of the facilities.

A plethora of local and international guidelines cover the design of tailings and water management systems and provide guidance to the designers. However, these do not cover every eventuality, design type or site condition that could be encountered. If in the opinion of the mining company these factors sit outside of the designer's experience, specialist advice should be sought.

As demonstrated in this chapter there are many elements and design parameters that feed into the design of the tailings and water management system. Many of these are developed at the completion of other portions of the study, such as mine production schedules, water demands and mine dewatering volumes. Sufficient time allowances should be provided to the designers of the tailings and water management system to complete their aspects of the study, which may impact the overall duration of the study.

Although there have been advances in tailings and water dam engineering design and construction over recent years, dam failures are still occurring at an alarming rate. Only when there is a shift in mindset away from cost minimisation toward a safety focus, as has occurred in most other areas of mining, will the rate of failures start to reduce. This shift in mindset is being endorsed by the design community, but the responsibility essentially lies with the owners of the facilities, the mining companies. The implementation of the *Global Industry Standard on Tailings Management* by the members of the ICMM is a step in the right direction. But to achieve the ultimate goal of zero harm to people and the environment with zero tolerance for human fatality across the broader mining industry it will require all mining companies, large and small, to adopt the same level of commitment to safe and sustainable management of tailings and water at their projects.

REFERENCES

ANCOLD (Australian National Committee on Large Dams), 2000. *Guidelines on Acceptable Flood Capacity for Dams* (ANCOLD: Hobart).

ANCOLD, 2012. *Guidelines on Consequence Category for Dams* (ANCOLD: Hobart).

ANCOLD, 2019a. *Guidelines on Tailings Dams: Planning, Design, Construction, Operation and Closure*, revision 1 (ANCOLD: Hobart).

ANCOLD, 2019b. *Guidelines for Design of Dams and Appurtenant Structures for Earthquakes* (ANCOLD: Hobart).

ANCOLD, 2020. *Guidelines for Geotechnical Investigations of Dams, Their Foundations and Appurtenant Structures* (ANCOLD: Hobart).

ANCOLD, 2022. *Guidelines on Risk Assessment*, July 2022 (ANCOLD: Hobart).

APEGBC (Association of Professional Engineers and Geoscientists of British Columbia), 2016. *Site characterization for Dam Foundations in BC*. Version 1.2 (APEGBC: Burnaby, BC).

Boughton, 2004. The Australian water balance model, *Environmental Modelling and Software*, 19(10):943–956.

Canadian Dam Association, 2021. *Tailings Dam Breach Analysis – Technical Bulletin* (Canadian Dam Association: Markham, Ontario).

Dam Safety New South Wales, 2022. Societal and individual risk rating methodology for *Dams Safety Act 2015* [online]. NSW Government Gazette, 18 March 2022. Available from: <https://gazette.legislation.nsw.gov.au/so/download.w3p?id=Gazette_2022_2022-113.pdf>.

DES (Department of Environment and Science, Queensland), 2016. *Manual for Assessing Consequence Categories and Hydraulic Performance of Structures*, Version 5.02 (Queensland Government: Brisbane).

DIIS (Department of Industry, Innovation and Science), 2016a. *Guide to Leading Practice Sustainable Development in Mining. Tailings Management* (Australian Government: Canberra).

DIIS, 2016b. *Guide to Leading Practice Sustainable Development in Mining. Water Stewardship* (Australian Government: Canberra).

Environment Canada, 2013. *Guidelines for the Assessment of Alternatives for Mine Waste Disposal* (Environment Canada: Gatineau, Quebec).

GTR, 2020. *Global Industry Standard on Tailings Management* [online]. Available from: <https://globaltailingsreview.org/wp-content/uploads/2020/08/global-industry-standard-on-tailings-management.pdf> (Global Tailings Review.org – co-convened by the International Council on Mining and Metals, United Nations Environment Programme and Principles for Responsible Investment), August 2020.

ICMM (International Council on Mining and Metals), 2021. *Tailings Management – Good Practice Guide* (ICMM: London).

ICOLD (International Commission on Large Dams), 2022. *Tailings Dam Design, Technology Update, Bulletin 181* (ICOLD: Paris).

ICOLD, in press. Bulletin 194, *Tailings Dam Safety* (ICOLD: Paris).

MAC (Mining Association of Canada), 2019. *Towards Sustainable Mining – Tailings Management Protocol* (MAC: Ottawa).

MAC, 2021a. *A Guide to the Management of Tailings Facilities,* Version 3.2, March 2021 (MAC: Ottawa).

MAC, 2021b. *Developing and Operation, Maintenance and Surveillance Manual for Tailings and Water Management Facilities* Version 2.1, March 2021 (MAC: Ottawa).

McLeod, H and Bjelkevik, A, 2017. Tailings dam design – technology update (ICOLD bulletin), in *Proceedings 85th Annual Meeting of International Commission on Large Dams* (National Committee on Large Dams: Prague).

Morrison, K F, 2022. *Tailings Management Handbook* (Society for Mining, Metallurgy and Exploration: Englewood, CO).

Pells, S and Fell, R, 2003. Damage and cracking of embankment dams by earthquake and implications for internal erosion and piping, in *Proceedings ICOLD conference* (ICOLD: Montreal).

Silva-Tulla, F, Lambe, T W and Marr, A W, 2008. Probability and risk of slope failure, *Journal of Geotechnical and Geoenvironmental Engineering,* 134(12):1691–1699.

Swaisgood, J R, 2003. Embankment dam deformations caused by earthquakes, in *Proceedings Pacific Conference on Earthquake Engineering* (New Zealand Society for Earthquake Engineering and New Zealand Earthquake Commission: Christchurch).

8

PROJECT DEVELOPMENT PLAN

INTRODUCTION

This chapter provides guidance for project development planning activities and plan definition by identifying the connections between key elements required for project development. The aim is to improve project development performance for resource projects. This is achieved through the following project development activities:

- studies, their planning, management and controls, and role in project commitment
- project readiness and project execution
- operational readiness and production ramp up.

Improvement in the development performance of resource projects is a necessary response to the evolving stakeholder expectations and their less than compelling delivery record. The improvement process entails distilling lessons and insights from past projects and adapting/adopting to changes in technology, for example.

It is important to recall that projects are approved for development by a company's Board of Directors. This seemly obvious and innocuous statement masks what is often a complex and sometimes lengthy process for some organisations. In this process, a project is promoted and championed by the executive directors led by the chief executive officer, or other executive, with the non-executive directors adopting the role of constructive but sceptical inquisitors of the information provided by an executive team. At the heart of this process is the trust and respect between these two groups (Knight, 2016, p 11).

In this process a project receives approval where the non-executives accept that a project is viable, it will provide superior returns to shareholders on the basis of reasonable economic assumptions, it can be permitted and built as specified, that the latent risks are manageable, but also that the owner's management has within the ranks of the organisation the expertise and commitment to deliver the project as proposed (Knight, 2016, p 13). A project's Feasibility Study is the primary source of information and the preparation of a well-executed Feasibility Study is an important prerequisite for this approval.

Table 8.1 presents the project development plan study phase guide, with further details provided throughout this chapter.

TABLE 8.1

Project development plan study phase guide.

PROJECT DEVELOPMENT PLAN					
Activity	**Deliverable**	**Description**	**SS**	**PFS**	**FS**
Study plan	Study plan	Objective	✓	✓	✓
		Scope	✓	✓	✓
		Report – internal*	✓	✓	✓
		Report – external	✓	✓	✓
		Budget	✓	✓	✓
		Schedule	✓	✓	✓
		Resources	✓	✓	✓
		Communication		✓	✓
		Change management		✓	✓
		Document management			✓
		Peer review		✓	✓
	Project execution and operational readiness plans	Not all study plan elements will be addressed in the Scoping Study phase, except where they are material to the project phase.		✓	✓

SS = Scoping Study; PFS = Pre-Feasibility Study; FS = Feasibility Study.

* It is important to note that 'Report – internal' refers to the elements in the report table of contents shown in Table 8.4.

STUDY TEAM

The owner's capacity to lead and manage the development of this strategy as part of a study is important. How this is achieved needs careful consideration. A common model is to manage studies via an owner's study team, who oversee and coordinate the assignment and management of study scope using specialists to complete the work. These specialists may include consultants and groups within an owner's organisation. The owner's study team establishment can respond to an owner's circumstances.

A junior or exploration company may rely on a small group of independent consultants with the support of the company's managing director to constitute a study leadership team. Whereas a mid-tier, or larger mining company may have staff it can assign to the owner's team. A mix between these two models is also common. An alternative model is handing completion of the study over to a large integrated consultancy or engineering firm. This approach risks the owner losing control of its ability to influence the project's identity in the study phase. Engineering firms, also, may have a conflict of interest when preparing studies, particularly a Feasibility Study, because:

- their motivation is winning work to build the project (McCarthy, 2016, p 26)
- their opportunity for making money in this part of a project life cycle is better
- they have the potential for greater fees compared with the study work.

McCarthy (2016) provides wide-ranging and practical commentary around how the types and level of a study (eg Scoping, Pre-Feasibility and Feasibility Studies) affect the organisation of studies, decisions on tendering out of study work, and the selection of the study manager and study management team.

The primacy of a well-prepared study, and by extension its preceding phases, as the foundation for the resources industry to make prudent decisions, renew itself and grow, is either alluded to or specifically captured in many of the papers referenced in this chapter. Practical lessons for preparing feasibility studies specifically are shown in Table 8.2.

A factor in the success of each item listed in Table 8.2 is the owner's preparation; in other words, its 'readiness' to undertake the work required. The importance of this preparation increases as work progress into construction and operations. For studies, this preparation is in the hands of the owner's study team and the resources available to them. Establishment of an owner's team for a study needs to consider:

- alignment with the owner's strategy and vision for a project
- consultants and specialists engaged to undertake specific roles
- competency of independent peer-review panels assembled to review the project's development, providing independent assessment to the owner, and support and guidance to the owner's team
- the financial backing and time to complete the work to the standards they require.

STUDY PLAN

The development plan for studies and their resourcing depends on which of the study phases is being executed. Figure 1.2 in Chapter 1 illustrates the components defining the basis for

TABLE 8.2

List of lessons learned preparing feasibility studies (after Shillabeer, 2001, p 440).

NO	LESSON
1	Don't begin the Feasibility Study until the project scope is defined and the economics have been optimised.
2	Work constantly to enhance the credibility of the Feasibility Study. Don't skimp on expertise.
3	Reserves should be validated before the Feasibility Study is well advanced.
4	Minimise the work done outside the task force.
5	Consider the personalities of the individuals involved on both 'sides' before agreeing to a pepper and salt arrangement, ie one where the owner's staff are seconded into a feasibility team managed and mainly staffed by an engineering contractor.
6	Produce and distribute the project description, summary execution plan, summary procedure manual and design criteria as early as possible. Keep these documents current and enforce them.
7	Establish an economic model early and update it. Use it to 'push back' against cost increases that are not accompanied by schedule or revenue enhancements.
8	The study manager should define at the outset and enforce his or her requirement for design criteria.
9	Use process flow diagrams in all areas where there is movement of materials or people.
10	Construct the project-wide mass and water balances and periodically update and elaborate them.
11	Leave plenty of time to review and check results.
12	Use only proven and familiar software.
13	Involve the owner's representative in informal initial design discussions before work begins, even when a study is completed 'in house', to ensure that polices are respected and the views of the business's operational group are respected.
14	Use milestones to record progress and challenge people. Ensure that everyone knows about the current milestones.
15	Establish at the beginning with the owner's representative the general policy towards change.
16	Plan from the beginning to be technically audited and arrange for it to be done during the study.

studies in each work phase. The matrix shows the differences between the level of accuracy of estimates expected in these studies. The preparation of a study plan provides a study with the best possible chance of success with the study phase they might be attempting. This is sometimes referred to as a 'study charter'. Its role is to provide a central point of reference for the following:

- scope definition
- standards of work
- allocation of responsibilities
- estimates of the study's time frame and its costs.

Study objectives

The aim of the study process is to develop a project sufficiently to enable a decision to proceed, or not, with its execution. Therefore, it is necessary for studies to provide an outline of the project's execution model and for the establishment of the operations, and alignment with the owner's corporate strategy.

Scope definition

Establishing a study scope of work is a foundation task for each study. Its purpose is linking the work of the study to corporate strategy and its objectives. This entails discussion of the following key topics:

- scope definition, including development and endorsement of a work breakdown structure (WBS) and capital and operating costs – addressed in Chapter 9
- identification and documentation of risks and opportunities for the project – addressed in Chapter 11
- future work, its definition and planning – this chapter.

Irrespective of the level of study being attempted it is essential this is prefaced by a clear statement of the study's purpose. This is essential for the study charter, the definition of the study's scope of work and its boundaries, the links to corporate strategy (McCarthy, 2016, p 21) and its objectives, and the establishment of clear accountabilities for study delivery. A pragmatic approach to this process is important. This provides the best opportunity for alignment of all key stakeholders and executive endorsement before investing time and money commencing work advancing the development of a study for a project and preparation of its business case for consideration by a Board of Directors (Singh, Harper and Burgio, 2023, p 85). Alternatively, if a project does not fit the owner's strategy or its capacity for its completion, this is flagged early and the study suspended or discontinued, eg the capital investment exceeds an owner's capabilities (McCarthy, 2016, p 21).

Irrespective of the level of study, the preparation of a forward (future) workplan covering tasks identified immediately following a study is desirable. These tasks address any further validation of assumptions, gaps in knowledge, risk prevention or mitigation activities, and opportunity management. Where possible this plan needs to cover task time frames, resources and costs for completion. Typically, each study chapter should include discussion of plans for future work pertaining to the chapter.

Internal study report

A useful bookend to the definition of the study's scope is the early preparation of a table of contents for the study. Where the scope addresses what the study covers, a table of contents provides an outline of the shape and form of what the study will look like as a delivered product. The early preparation of a table of contents provides the study team with something they can reference as part of discussion and negotiations directed at ensuring study scope has not missed any material components or has unnecessary repetition. Additionally, it provides a tangible product for presentation and discussion with those external to the study to quickly get an idea of what to expect a study to look like and against which they can seek any clarification. A draft table of contents is a useful inclusion to a study plan as an appendix.

The structure and format of the table of contents supports access to project information by the project's stakeholders. These stakeholders may include:

- owner's Board of Directors
- government departments and regulators across national, state and district jurisdictions
- community – state, district and local
- the investment community with whom an owner may need to engage to support funding of the project's development
- people working with the owner with the goal of establishing a sustainable business based on the future operation.

External study report

This is not to be confused with the prescribed formats for Public Reporting of studies for resource projects and their presentation to government. Amongst the most prominent is Canada's National Instrument (NI) 43-101 Standards of Disclosure for Resource Projects, NI Technical Reports and related consequential amendments. Some countries have prescribed tables of contents for studies submitted to government. These studies are often created from, or lean heavy on, work generated by a studies workflow described in this chapter. Large and mid-tier mining organisations usually have their own standards for the preparation of studies aligned with their own requirements and matched to commonly recognised study phases. As outlined in this chapter's introduction, these phases each have a purpose. In some cases, their end use is to support the commercial needs of the business. For example, in the case of an exploration focused company, or a mining junior, their capacity to attract funding for the project, or in the case of a larger business to determine what projects from within their project pipeline are funded, paused or disposed of through sale or relinquishment.

In most cases the description and reporting of studies for resource projects have strong connections with national guidelines and codes for the definition of Mineral Resources and Ore Reserves. For example, The Australasian Code for Reporting of Exploration Results, Mineral Resources and Ore Reserves (the 'JORC Code', 2012), which is a professional code of practice that sets minimum standards for Public Reporting of mineral Exploration Results, Mineral Resources and Ore Reserves. The JORC Code has been incorporated in the Listing Rules of the Australian and New Zealand Stock Exchanges, making compliance mandatory for listing public companies in Australia and New Zealand.

Budget and schedule

Estimates of time and cost are as relevant to studies as they are for most other aspects of project development; businesses have an inherent need for knowledge of both. McCarthy refers to the project management triangle (McCarthy, 2016, p 27): a study can be of high quality, completed quickly, or of low costs. A study can be conducted anywhere within a triangle defined by these three points. The inference being a high-quality study is unlikely to be quick and low cost. To this point an owner and their study team can and will likely make their own trade-off decisions regarding a study's preparation. This is likely to fit the owner's business needs and constraints regarding the study time frame and cost. These will depend on the type and size of the study, and approach to its preparation and the allocation of responsibilities for its completion. The challenge is always for a study's estimates for its time frame and costs to be realistic and include capacity to genuinely deal with the limits of their knowledge at the start of a study.

Schedules

The preparation of schedules for resource project studies, while useful, needs to be mindful that the inherent objectives of these studies is to optimise the project's configuration for construction and operation to deliver the maximum possible benefit for the technical and commercial risks involved when expressed in terms of the corporate goals of the owner (McCarthy, 2016, p 21).

Technical experts reviewing a completed study are often interested in the ideas rejected, and the process used to discard them. The visibility and access to this information allows them to assess the rigour and completeness of the work undertaken by a study team. This approach aligns well with the JORC Code concept of 'if not, why not'.

Some flexibility or allowance in time and funding needs to be available for this process. It is the uncertainty of this exploration of ideas and alternatives that makes the preparation of the study phases for minerals projects different from the more defined aspects of the following work, for example, detail design and engineering as part of its execution. The study sequence described in this handbook matches this approach. However, often against the background a misplaced sense of urgency and/or a requirement for compressing the time frame, an owner may wish to bypass a study phase. Most often this is the Pre-Feasibility Study. This is ironic, because this is the phase where most of the big decisions that shape the form of a project are made. This approach to completing the study process for a project risks the presentation of a poorly conceived project.

The time frame for project development varies as a function of a project's maturity, its scale and complexity, the continuity and integrity of proceeding work, and the requirements of the owner's needs and their familiarity with the commodity and project's setting. In general terms project development timeline components recognised in studies could include:

- drilling and assay programs
- test work to determine physical and metallurgical properties
- study phases
- permitting and approvals required by governments
- project execution phases.

These programs may not be within the direct control of the study team. For example:

- drilling for Mineral Resource definition and geotechnical information for infrastructure, facilities and for mining evaluation and designs
- metallurgical and material properties test work and assessment, including those needed for tailings storage facilities.

These programs may precede a study or are completed in parallel. Whatever the case, the study is their customer and needs some ability to influence the scope and scheduling of the work. Without it, the integrity of a study is put at risk by denying a study timely access to necessary information from these programs.

Useful study milestones to identify as part of a study schedule could mark completion of:

- a Mineral Resource estimate update for the project
- metallurgical and physical properties test work used to support the Feasibility Study process engineering, flow sheet selection and plant capacity settings
- an initial Ore Reserve estimate for the project and an accompanying mine design and production schedule
- infrastructure and non-processing facility configuration design and preliminary engineering
- preliminary project implementation plans
- an assessment of risks and opportunities for project delivery and operations
- Feasibility Study project schedule, revenue and cost estimates.

Some of the above milestones may have links with predetermined hold points for project outcomes and the independent peer-review process discussed later in this chapter. These hold points provide a mechanism for project oversight by an owner's executive and Board of Directors up to completion of a study phase. The results from this process range from their endorsement to direction regarding subsequent work.

Study cost

Estimate approaches for study costs vary. At one end of the spectrum is the use of thoughtful and informed guesstimates to generate a study cost estimate. At the other the end, most of the cost estimate is supported by a request for proposals and quotations from competent and experienced professionals. These requests may include a well-defined scope of work with clear deliverables. The larger and more significant a study, the more the owner pushes it towards a more comprehensive approach to study estimate preparation. This process has its own demands for time to prepare scopes of work, engage with potential expert providers and to evaluate their responses to finalise a study plan. These costs are easy to underestimate or overlook.

Study management and controls

Beyond the need for clarity in preparation of studies discussed earlier, they require effective allocation of responsibilities, internal communication and coordination, change management, knowledge management (ie document management) and independent peer-review processes. All these processes relate to one another and form the machinery of a study.

Allocating responsibilities

The allocation of accountabilities within a study team is essential for their preparation and to limit any tendencies for duplication or unnecessary repetition. The RACI (responsible, accountable, consulted, informed) table (Table 8.3) provides an outline of the accountabilities and responsibilities associated with a Feasibility Study as an example.

A project owner should guard against overconfidence in their own ability and the adequacy of their team's skills in carrying out a study. Both have capacity to impair the results of the study produced, eg underestimates of capital cost estimate, ramp-up time frames and working capital requirements. They need to recognise that study work entails a related but

TABLE 8.3

RACI for a process management study.

NO	STUDY – PROCESS MANAGEMENT	EXECUTIVE SPONSOR	STUDY MANAGER	CHAPTER LEADERS	CONSULTANTS	INDEPENDENT REVIEWERS
1	Scope and strategy management	**A**	**R**	C	I	C
2	Preparation of section/subsection scopes of work and consultant selection	C	**A**	**R**	I	C
3	Tasking and tracking work teams at section and subsection level to complete agreed scopes of work	I	**A**	**R**	C	I
4	Risk and opportunity management; assessment monitoring and tracking as appropriate	C	**A**	**R**	C	C
5	Independent review processes	**A**	**R**	C	I	-
6	Document control and definition of work breakdown structure	C	**A**	**R**	C	I

RACI = **R**esponsible – owns the work; **A**ccountable – must sign-off/approve the work; **C**onsulted – needs to be consulted as part of work preparation; **I**nformed – must be informed of the results of the work but does not need consultation for its preparation.

different set of skills to those required in project construction and operations (Kear, 2004, p 117–118). These skills are not simply acquired by a change in role title.

Typically, the study team's structure is aligned with the study's table of contents, particularly in later study phases. This approach provides clear lines of accountability for leadership, management and delivery of study chapters and their sections. Table 8.4 shows an example of this approach. In this model, each chapter may have its own summary and a separate volume of appendices, which contains the substantive detail of work referenced in the chapter. The chapter summaries provide the primary content for preparation of a study's executive summary.

TABLE 8.4

Possible study chapters and specialist responsibilities.

CHAPTER NUMBER	STUDY CHAPTERS	POTENTIAL RESPONSIBILITY
0	**Executive summary**	Owner's study manager
1	**Introduction** Consolidation of generic reference information (so there is no need for its repetition in other chapters), eg location, project or property ownership, climate, geopolitics.	Owner
2	**Legal, approvals and permits** • Description of the relevant legal and regulatory environment	
3	**Environment, social and permitting** • External stakeholder relations (state and national government / regulator focused) • Environmental and social studies • Environmental and community relations issues and their management • Initiatives and programs supporting social licence • Basis of the workforce deployment and its structure in these areas • Establishment of a register of approval and permits, their status and requirements	

TABLE 8.4 CONT...

CHAPTER NUMBER	STUDY CHAPTERS	POTENTIAL RESPONSIBILITY
4	**Human resources** Description and discussion of the scope and objectives for: • Workplace health and safety • Human resources / people and culture • Workplace health and safety, includes on-site medical and emergency response capability, etc • Site services infrastructure and facilities, eg accommodation and offices, etc • Site access and security • Supply – purchasing, logistics and warehouse • Information and communications technology • Basis of the workforce deployment and its structure for mineral processing operations • General and business management, includes the off-site offices, general manger operation's office and commercial functions • Basis of the workforce deployment and its structure in these areas	Owner supported by a consultant(s)
5	**Geology and Mineral Resources** • Geology • Resource model • Quality assurance and quality control	Consultant
6	**Mining and Ore Reserves** • Method selection • Mine optimisation • Mine geotechnical • Hydrogeology; site general including water supply, and mining • Mine design and engineering • Life-of-mine schedule • Mine operations description • Equipment selection and fleet estimation • Basis of the workforce deployment and its structure for mining operations • Ore Reserves statement	Consultant
7	**Metallurgy** • Investigations, test work and test work interpretation • Flow sheet selection • Process plant design criteria	Consultant
8	**Engineering** Process plant and infrastructure: • Infrastructure – power and water supplies, access to site and around the site • Facilities – accommodation, office, workshops, warehouses, laydown areas, site entry / security points, waste (rubbish) disposal, etc Site geotechnical Tailings: • Investigations, site assessments, test work and its interpretation • Tailing disposal methodology evaluation and selection • Tailings disposal design and engineering Water and sediment management: • Hydrology and water management • Run-off and sediment controls • Description / definition of the project's water balance	Consultant

TABLE 8.4 CONT...

CHAPTER NUMBER	STUDY CHAPTERS	POTENTIAL RESPONSIBILITY
9	**Implementation strategy** • Project development • Implementation and contracting strategy • Project execution plan and its organisational structure • Operational readiness and organisational structure for operations • Basis of the owner's costs for project execution and operational readiness	Owner supported by a consultant(s)
10	**Marketing** • Market locations and capacity • Future outlook • Long-term pricing • Feed • Product • Treatment and refining charges • Freight	Consultant
11	**Capital cost** • Capital cost – direct and indirect • Contingency and contingency modelling • Sustaining capital	Consultant
12	**Operating cost** • Production schedule • Reagents and consumables • Power • General and administration • Operations and maintenance	Consultant
13	**Economic evaluation** • Production and revenue summary • Financial analysis • Financing options • Sensitivity analysis • Multi criteria analysis for options	Owner
14	**Risk** • Enterprise • Climate • Technical • ESG and permitting	Owner and consultants
15	**Path forward** • Forward work plan • Budget • Resources • Schedule	Owner

Internal communication

The proper leadership, coordination and organisation of studies is important – teamwork matters. This depends on open and honest communication between all areas of a study. This is difficult to achieve using a siloed approach. Regular collaborative communication, which brings together all the study components, is more challenging to manage in a time-efficient way. However, when done effectively it provides the leaders of study areas with the situational awareness to manage their work and contribute to a study's completions, without gaps and unnecessary overlaps. Usual tools for this process are meetings, their minutes and progress reporting. These need to be timely and fit for purpose, and forward looking.

External communication

A project owner's stock exchange listing requirements and commercial circumstances drive its use of information generated by its studies work; for example, an exchange's general notification obligations regarding material changes in information concerning the owner's business from that previously provided (Australian Securities Exchange [ASX] Operating Rules, Guidance Note 8). Or it may simply be the case of an owner feeling a need to maintain a 'news flow' to the market in line with previous guidance. An owner's funding needs may require it to engage with potential third parties regarding sources of debt or equity funding associated with a project. This process may entail preparation of an independent technical expert's (ITE) report based on their review of the completed study work and its supporting information. Both the ITE report, and for example the Feasibility Study, could be provided to third parties considering contributing to project funding.

Change management

Studies for the development of resource projects inherently entail the creation and organisation of knowledge pertaining to the project. Implicit in this process is the evolution of ideas and concepts based on this knowledge, which requires discipline around change management. As a resource project moves through the study phases, it is important the decisions material to the definition of the project's development are identified and captured. A decision register is commonly used for this purpose. It records these decisions, any background, process applied, references supporting information covering alternatives, the role accountable for the decision and date it was made. The materiality of these decisions guides which role owns it. The usefulness of a process of this type is that it limits churn and drift in studies. It also makes progress visible along with the accountability for decisions.

Document management

Any significant study will require some level of document management and control to deal with the intellectual property generated by the study's activities. Their purpose is simply to provide clarity around version control and its alignment with ownership and accountability for work completed. Commonly these processes are organised around a document numbering system linking documents to areas of scope covered by the WBS. Whatever the approach adopted by a project owner, a document management system should provide it with a single comprehensive source of information pertaining to study work. Many of the documents compiled in a study's document management system will likely appear as appendices to individual study chapters.

Peer review

Independent reviews at appropriate points in any substantial study are an integral and important component of the proposed study process. These review points will be defined in consultation with the owner. Their purpose is to support the study team to maintain objectivity and ensure that appropriate rigour is applied. The review process is intended as an interactive and constructive one, providing independent review and constructive feedback to the study team. Wherever possible, issues are discussed and addressed with the study team before reporting to the owner. Emphasis is on investigation and verification of processes, their supporting facts, and validation of assumptions. It is not on duplication of project work. This type of peer-review process performs best where the reviewers are competent and independent of the study team, hence the use of terms like 'competent independent review' or 'independent peer review'. Importantly they are termed reviews rather than audits and make a clear distinction from any tendency to referencing 'checklists' or 'tick a box' approaches. Their effectiveness depends on the quality of the engagement and discussion between the study team and the reviewers.

PROJECT EXECUTION

The definition of project readiness for execution (construction) and its role in the performance of resource projects is something not clearly addressed in most discussions around the performance of resource projects, and how it links with the phases on either side. The coverage of project readiness available in the literature reviewed for this chapter was limited at best. What is clear in this literature is the importance of readiness as a broad concept generally applicable to many aspects of a successfully functioning business. In this context readiness is flagged as the missing, or overlooked, link between strategy, the intent and decision to take a course of action, and the action's actual execution (Rey, 2019). Readiness is the enabler of an organisation's ability to 'walk the talk', which sets it up for success! Accepting this proposition leads to the conclusion that readiness for project execution on the part of an owner likely plays a role in the successful completion of a project's execution phase.

The project execution plan (PEP) documents the scope of the execution methodology, policies, plans and procedures for the delivery of a project. The PEP aims to ensure a technically successful project is achieved safely, within the terms of its environmental approvals, and its approved budget and schedule. The PEP features in the preparation of project studies and continues to be developed during the detailed design phase and will be updated as required. In addition to the study plan elements discussed above, the topics covered in a PEP could also include:

- Project execution strategy
- Project deliverables
- Project work breakdown structure
- Project management
  - Project personnel requirements
  - Project controls
  - Project safety
  - Contracting strategy
  - Budget (capital expenditure [CAPEX])
  - Schedule
- Engineering management
  - Engineering services
  - Standards and codes
  - Design health safety environment integration

- Procurement and contracts management
  - Procurement
  - Contracts management
- Construction management
  - Construction sequence
  - Commissioning
  - Operational readiness (see below)
  - Handover and transition to operations
- Transport and logistics strategy
  - Ports
  - Road transport
  - Lifting
  - Laydowns
  - Vehicle movements.

The handover process for a project from study to project phase is done at, or around, the time of the decision to proceed with the project. Those who led the study phase step back to make way for those focusing on leading construction working to a PEP. The construction and execution tasks require a different skill set. The problem space is all about delivering the project's physicals: the scope, cost and time to complete this task. It is this problem space that is at the heart of the PEP, the delivery of the project's physical components, and managing its construction, including:

- scoping, engineering, design and construction of infrastructure and facilities
- specification, procurement and delivery of these items and the commissioning of equipment and facilities required for mining and processing operations.

This is the realm of the project manager and the project team, whose focus is very different from the general manager and the operating team. The project team has a short-term involvement, which continues until plant commissioning is complete and project custody transitions to the general manager and the operations team. The general manager and operating team are there for the long term, their focus is on the operation of the project's assets in accordance with the owner's operating strategy and plans. That is, the operation's production. The operation's mining team will likely lead this process off through their work in pre (plant) production pit and/or mine development.

Successful projects are delivered by teams of aligned and committed people, properly led and organised, with the skills and experience to maximise level of assurance for delivery: on time, on budget and meeting the performance criteria set in the Feasibility Study (Knight, 2016, p 15). This statement presupposes and requires a project owner to have completed a requisite level of project readiness and compatible PEP.

An owner whose project readiness is not consistent with a project's PEP might find they struggle to support their project's development competently and confidently. This may have implications for the time and cost to complete the project.

OPERATIONAL READINESS

For almost all new operations, getting away to 'a good start' is universally assumed to be true in the base case of the Feasibility Study (Pratt, 2023, p 157). In this context, 'good' means achieving all the targets set for its first 12 month to two years. Achieving these targets takes place against the background of the owner's operating team coming to grips with:

- new equipment and how they are going to use it
- the processes for running operations to the satisfaction of the owner
- the project location
- the deposit (possibly the first time it has been mined and processed)
- the relationships built within and outside their team that are essential for its performance
- inheriting the relationships, good and bad, established by the project team during the construction period – that is, the project's identity and reputation with its stakeholders.

The complexity and challenges of this task are influenced by the familiarity of the business owner behind the project with the scale, type of operation and its location. This familiarity is the starting point for their operational readiness.

The owner's role in the project's construction is also a factor in operational readiness. This is because it is a related and subsequent stage in the development of business (operating) management systems. Operational readiness entails the evolution of some of the processes discussed earlier in the chapter for project readiness, management plans and their procedures, and the addition of those specific for the operations phase.

For the purposes of this chapter, the term 'operational readiness' describes the work required to ensure that the owner or operator of a project or asset under construction is equipped to take delivery of the asset at handover from the construction phase. Implicit in this description is that, as the operator, they can accept responsibility for the asset and operate it in a way that is consistent with their business needs and chosen risk appetite (Pratt, 2023, p 158).

To match this description requires:

- accessing and developing the skills and capabilities (knowledge) required to support the operation, including its maintenance – that is, the recruitment of operations and maintenance personnel and their initial training
- labour resources planning, allocating accountability and defining communication within the organisation to ensure an operation can achieve its operational and business objectives, including transition of personnel from project to operations, inclusive of contractors, while optimising local employment
- defining the production ramp-up program and setting performance targets
- identifying and developing the protocols and processes required to commence operations and equipment maintenance, including risk management, and gaining timely access to project documents and vendor information
- consultation with stakeholders, community and regulatory authorities, to build effective relationships and provide appropriate information to establish and maintain their support for the operation
- introduction of equipment to the mine and its ramp-up program.

Preparation of an operational readiness plan (ORP) entails:

- understanding the business's capacity to support the new operation against the background of its business management framework
- evolving the business management systems used to support the project's construction phase and adding those systems needed for the operating phase.

This work is required to ensure a smooth transition from the project phase to initial operation and to sustain the operation's future performance. The ORP must address what the operation needs to get right from the start of operations to meet its best-case scenario (Deloitte, 2013). The key questions for an OPR to address include:

- What objectives does it need to pursue to achieve the targeted critical outcomes?
- What are the tasks and process areas that support these objectives?
- Who are the operation's business partners?
- What are their roles and the nature of their relationship?
- How is ORP progress measured and communicated?

The owner's business management framework is a key part of the foundation for operations readiness, as it is for their approach to 'project readiness'.

The description of operational readiness at the beginning of this section is theoretical. In practice, a future operation needs to come up with a description for operational readiness that is meaningful to its circumstances and its people. This is important to building their commitment to operational readiness preparations. The intent of operational readiness is to avoid an operation falling victim to the fallacy that the personnel who operate and maintain the business will build the processes they need as they go (DiStefano, Gotetz and Storino, 2015).

The requirement for operational readiness has clear implications for the preparation of the project's capital costs. This work needs its own line item in the WBS. The work requires people (ie the operations team), takes time and needs money for its completion.

Many guidelines and standards for mining project studies require statement of an operation's operating and maintenance philosophy, organisational structures, mobilisation schedule for operations personnel, etc. If this is merely a purposeful description of requirements resolved in spreadsheets of costs, it is easy to overlook the resources, time demands and scale of the actual process of recruitment. It also does not address questions of what personnel are required to do when they arrive, the time and resources they will need for operational readiness, and their estimated costs.

The topics covered in an ORP could include:

- Introduction
    - Project overview
    - Project schedule
    - Operational readiness, and its purpose
- ORP approach
    - Link with the operating risk register
    - Assumptions, limitations and exclusions
    - Definition of the operational readiness process applied across all functions of the operation
    - Leadership of operational readiness
    - Roles and responsibilities
    - Statutory functions and compliance roles
    - Operational readiness personnel requirements
- Establishment of ORP activities
- Tracking and monitoring the work
- Change and communication management

- Document management
- Operational readiness resources and budget.

CONCLUSIONS

The start of this chapter claimed the successful development of resource projects required the satisfactory execution of the following:

- *studies* – culminating in the completion of a Feasibility Study that provides the basis for a project's implementation
- *project readiness* – the owner's readiness to proceed with the execution of the project's construction
- *project execution* – delivery and construction of the project's physical operating assets
- *operational readiness* – the owner's readiness to commence operating the project's assets
- *ramp up* – scaling the project up to its operating performance target.

The discussion of each of these steps in this chapter provides a starting point for understanding the factors on which successful resource project developments depend. The chapter comments on balancing consideration of a focus on tangible elements of the resource project, ie building the physical asset, and intangible aspects of project development, the human factors. What is important to both is teamwork; people, the processes they apply and the relationships they build, ie highly effective teams.

The creation of highly effective teams, those capable of the requisite teamwork, depends on consideration of the human factors in play within teams: trust, respect, strong leadership, complementary skills, open communication, lack of selfishness and unity (Ebbles and Pratt, 2012, p 200). Demonstrable intelligence by teams who have custody of a project during its development depends on their ability to understand a project within its host environment, to adapt the owner's strategy and to adopt tactics for project delivery in a way that meets the owner's requirements.

In most cases the teams addressing each of these steps for project development are not the same. Adjustment to their composition is required to achieve the objectives for each of these steps, and to match skills sets required. As emphasised in this chapter, successful project development requires each step to have an effective engagement with its preceding and following step. This requires strategic and executive leadership and teamwork. Project development depends on teamwork. Excellence in one or just a few areas will not successfully bring on the development of a project. In this respect, project development is like many areas of human endeavour, as said by basketballer Michael Jordan, 'talent wins games, but teamwork and intelligence win championships'.

ACKNOWLEDGEMENTS

The author would like to acknowledge the contribution of colleagues and consulting clients who have been part of the discussion and debate that informs this chapter. However, any errors or misconceptions are the author's alone.

REFERENCES

Deloitte LLP, 2013. Operations go-live! Mastering the people side of operational readiness [online]. Available from: <https://www2.deloitte.com/content/dam/Deloitte/global/documents/gx-icp-operation-go-live.pdf>.

DiStefano, B, Gotetz, W and Storino, B, 2015. Operational readiness: bridging the gap between construction and operations for new capital assets [online]. Available from: <https://reliabilityweb.com/>.

Ebbles, A M and Pratt, D, 2012. A great project team – bring together location and culture, in *Proceedings Project Evaluation 2012*, pp 197–202 (The Australasian Institute of Mining and Metallurgy: Melbourne).

JORC, 2012. Australasian Code for Reporting of Exploration Results, Mineral Resources and Ore Reserves (The JORC Code) [online]. Available from: <http://www.jorc.org> (The Joint Ore Reserves Committee of The Australasian Institute of Mining and Metallurgy, Australian Institute of Geoscientists and Minerals Council of Australia). Accessed 16 April 2023.

Kear, R M, 2004. Mine project life cycle, in *Proceedings MassMin 2004* (eds: A Karzulovic and M A Alfaro), pp 117–120 (Instituto de Ingenieros de Chile: Santiago).

Knight, R, 2016. Project evaluation – a non-executive director's perspective, in *Proceedings Project Evaluation 2016*, pp 11–19 (The Australasian Institute of Mining and Metallurgy: Melbourne).

McCarthy, P, 2016. Making studies work – organisation, communication and management, in *Proceedings Project Evaluation 2016*, pp 20–30 (The Australasian Institute of Mining and Metallurgy: Melbourne).

Pratt, A G L, 2023. Operational readiness – what is it and how to get it done?, in *Proceedings Underground Operators Conference 2023*, pp 156–175 (The Australasian Institute of Mining and Metallurgy: Melbourne).

Rey, S, 2019. Readiness: the missing key between strategy and execution [online]. Available from: <https://trainingindustry.com/articles/strategy-alignment-and-planning/readiness-the-missing-key-between-strategy-and-execution/>. Accessed 17 May 2022.

Shillabeer, J H, 2001. Lessons learned preparing mining feasibility studies, in *Mineral Resource and Ore Reserve Estimation – The AusIMM Guide to Good Practice* (ed: AC Edwards), pp 435–440 (The Australasian Institute of Mining and Metallurgy: Melbourne).

Singh, U, Harper, P and Burgio, N, 2023. Planning for underground mining at Savage River Iron Ore Mine, in *Proceedings Underground Operators Conference 2023*, pp 84–93 (The Australasian Institute of Mining and Metallurgy: Melbourne).

COST ESTIMATION

9

INTRODUCTION

Cost estimation is an essential part of Scoping, Pre-Feasibility and Feasibility Studies as a key input to financial evaluation of a project. Each phase of study requires a different level of detail and accuracy in cost estimation. Cost estimates developed at each study phase produce different and increasing confidence levels of accuracy that align with the decisions required to progress to the next phase of the project.

This chapter provides guidance on best practice in cost estimation of mining projects through different phases of development from Scoping through to Feasibility. This chapter is prepared in a broad context and is not specific to any particular commodity or market; however, it aims to provide the reader with some key concepts to consider when preparing a mining project study. It also includes guidance for best-practice modelling.

Further detail about cost estimation techniques, data sources and estimation approaches can be found in the Australasian Institute of Mining and Metallurgy (AusIMM) Monograph 27 – *Cost Estimation Handbook* (2013). This chapter has drawn some content from that handbook.

The requirements of a Scoping Study, Pre-Feasibility Study and Feasibility Study are presented in the context of economic evaluation in Chapter 10, in the section titled 'Understanding the confidence level of the study', but largely the principles and recommendations provided in this chapter apply across all three study levels, except where stated otherwise.

Figure 9.1 is an extract from the study processes matrix, showing the differences between the level of confidence for cost estimation for each of a Scoping Study, Pre-Feasibility Study and Feasibility Study. A full copy of the matrix is provided as Figure 1.2 in Chapter 1.

	SCOPING STUDY	PRE-FEASIBILITY STUDY (PFS)	FEASIBILITY STUDY (FS)
CAPITAL COST ESTIMATE	Accuracy -25%+50%, Contingency ±30%	Accuracy -15%+25%, Contingency 15–30%	Accuracy -10%+15%, Contingency 10–15%
OPERATING COST ESTIMATE	Accuracy -25%+50%, Contingency ±25%	Accuracy -15%+25%, Contingency ±15%	Accuracy -10%+15%, Contingency ±10%

FIG 9.1 – Cost estimation levels of confidence.

AusIMM's Monograph 27 – *Cost Estimation Handbook* presents a table showing generic study classifications often used in the minerals and other industries. This table is presented here (see Table 9.1) as a cross-reference between these commonly used terms and the elements presented in this document.

It is important to note that study cost estimation is subject to various uncertainties and risks, including commodity price fluctuations, regulatory changes and unforeseen technical challenges. Therefore, sensitivity analysis and risk assessment are typically conducted to understand the potential impact of these factors on the project's financial viability. Mining companies often engage qualified professionals and consultants with experience in cost estimation to ensure accuracy and reliability throughout the study phases.

In each study phase, cost estimation involves iteration, sensitivity analysis and risk assessment to ensure a robust evaluation of the project's financial viability and to identify potential cost-saving measures or risks that may affect profitability.

TABLE 9.1

Generic study classification guide (Monograph 27 – *Cost Estimation Handbook*).

TERMINOLOGY USED IN THIS HANDBOOK		SCOPING STUDY – PHASE 1	PRE-FEASIBILITY STUDY – PHASE 2	FEASIBILITY STUDY – PHASE 3
Front end loading		FEL 1	FEL 2	FEL 3
Different titles that may be used to describe this level of study	Conceptual	Concept	Preliminary feasibility	Final feasibility
	Opportunity assessment	Order of magnitude (OOM)		Basic engineering
		Identification phase	Selection phase	Definition phase
	Screening	Scoping		'Bankable' feasibility
	Scoping			Definitive feasibility
		Capacity factor	Equipment factor	Forced detail
		Preliminary evaluation	Intermediate economic study	
Estimate type (AACE)		Class 5	Class 4	Class 3
Expected accuracy range of capital cost	±35% to ±100% Typically ±50%	±30% to ±35%	±20% to ±25%	±10% to ±15%
Expected estimate contingency range	30% to 75%	20% to 35%	15% to 25%	10% to 15%
Level of definition (% of complete engineering	Minimal, generally based on other operations, or in-house 'database'	1–2% Basic general layouts	10–15% Preliminary take-offs	15–25% Detailed drawings and take-offs
Typical estimating methodologies	Capacity factored Parametric models, judgement or analogy Stochastic estimating methods, including cost-capacity curves, and various factors	Equipment factored or parametric models. Some 'first principles' estimating related to early scope definition	Semi-detailed unit costs, and more deterministic estimating methods Preliminary material take-offs (Some) budget pricing	More detailed unit costs and material take-offs Budget prices and vendor quotes Higher degree of deterministic estimating methods Line items, and forced detail where definition is lacking

FEL = front end loading; AACE = Association for the Advancement of Cost Engineering.

The study cost estimation process is typically conducted by a team of specialised consultants and experts in mining engineering, geology and project management. The process typically involves the use of specialised software and modelling tools to estimate costs and evaluate project economics.

MODELLING METHODOLOGIES AND TOOLS

In mineral project studies various cost modelling tools are available to assist in cost estimation processes.

These tools are designed to facilitate the calculation, analysis and projection of costs associated with mining projects or operations. Here are some commonly used cost modelling tools:

- *Spreadsheets* – Spreadsheets, such as Microsoft Excel or Google Sheets, are widely used for cost modelling. They provide a flexible platform for creating customised cost models and performing calculations. Spreadsheets allow users to input cost data, apply formulae, create scenarios and generate reports. They can be tailored to specific cost components, such as equipment, labour, materials and maintenance.
- *Cost estimation software* – There are specialised software applications available specifically for mining cost estimation. These software tools provide pre-built cost models, templates and libraries of cost data that can be customised to suit specific mining projects or operations. They often include features such as data integration, scenario analysis, sensitivity analysis and reporting capabilities. Examples of popular mining cost estimation software include MineRP, CostMine and RPMGlobal's XPAC Solutions.
- *Financial modelling software* – Financial modelling software, such as discounted cash flow analysis tools or project evaluation software, can also be utilised for cost estimation. These tools enable users to assess the financial viability of mining projects by considering cash flows, capital expenditures, operating costs and revenue projections. Financial modelling software often includes built-in mining-specific features and assumptions, such as production profiles, commodity prices and discount rates.
- *Simulation and optimisation tools* – Simulation and optimisation software allows for complex modelling of mining operations and cost estimation. These tools use algorithms and mathematical models to simulate various scenarios, optimise processes and estimate costs. They can help analyse production schedules, equipment utilisation, resource allocation and cost optimisation strategies. Examples of simulation and optimisation software used include Witness, Arena and PROMODEL.
- *Integrated mining software suites* – Integrated mining software suites provide comprehensive solutions that encompass various aspects of mining operations, including cost estimation. These suites often include modules or functionalities for cost modelling, budgeting, resource planning and financial analysis. They integrate data from multiple sources, such as geological models, production data and cost databases, to generate accurate cost estimates. Popular integrated mining software suites include software packages offered by mining technology companies like Hexagon Mining, Deswik and RPMGlobal.

When selecting cost modelling tools, it is important to consider factors such as user-friendliness, compatibility with existing systems, scalability, data integration capabilities and the specific requirements of the mining project or operation. Additionally, training and support options provided by the software vendors should be considered to ensure effective utilisation of the selected tool.

Scoping Studies

For Scoping Studies, which are preliminary assessments conducted to determine the potential viability of a mining project, it is important to use cost estimation techniques that are relatively quick and simple, and that require minimal data. These techniques provide rough order-of-magnitude estimates to guide early-stage decision-making. Here are some cost estimation techniques commonly used in mining Scoping Studies:

- *Analogous cost estimation* – This technique involves using cost data from similar mining projects that have been previously completed or are currently in operation. By comparing the characteristics and scale of the current project with historical data, you can estimate costs based on similarities. However, it is essential to ensure that the projects being used as analogues are reasonably comparable.
- *Factor estimation* – The factor estimation method uses cost factors (multiples) to estimate the overall project cost based on certain key parameters, such as production capacity, ore tonnage or mining method. These factors are derived from historical data and are applied to the specific project to arrive at an approximate cost estimate.
- *Parametric cost estimation* – Like factor estimation, the parametric cost estimation method uses equations or statistical models to estimate costs based on specific project parameters, such as the number of mining equipment units, processing plant size or infrastructure requirements.
- *Scalability index method* – This method uses scaling factors to adjust costs from a known reference point to the current project scale. For example, if you have cost data for a smaller mining operation, you can use scaling factors to estimate costs for a larger operation.
- *Cost capacity factors* – This technique estimates costs based on the production capacity of the mine. The costs are typically presented in cost per unit of production (eg cost per tonne of ore mined or processed).
- *Vendor quotes and budgetary quotes* – For certain equipment or infrastructure components, reaching out to vendors for preliminary quotes or using budgetary quotes can provide quick estimates for specific items in the project.
- *Historical cost data* – Utilise any available historical cost data from previous projects undertaken by the company or industry to inform the current cost estimates.
- *Expert judgement* – In some cases, experienced mining professionals may use their judgement and expertise to provide rough cost estimates based on their knowledge of similar projects and market conditions.

Mining Scoping Studies are early-stage assessments, and the cost estimates obtained from these techniques are meant to be indicative and preliminary. As the project advances through the study phases, more detailed and accurate cost estimation methods should be employed to refine the estimates further.

Pre-Feasibility Studies

For Pre-Feasibility Studies, more detailed and refined cost estimation techniques should be used compared to Scoping Studies. Pre-Feasibility Studies involve a more comprehensive evaluation of the mining project's technical and economic viability, and the accuracy of cost estimates becomes increasingly critical. Here are some cost estimation techniques commonly used in mining Pre-Feasibility Studies:

- *Definitive cost estimates* – As the Pre-Feasibility Study progresses, strive to transition from order-of-magnitude estimates used in Scoping Studies to definitive cost estimates. These estimates are based on detailed engineering designs, equipment specifications and vendor quotations. Definitive cost estimates provide higher accuracy but require more time and effort to develop.
- *Detailed quantity take-off* – Conduct a detailed quantity take-off, which involves itemising all the materials, labour, equipment and services required for the project based on the engineering design. This provides a foundation for accurate cost estimation.
- *Bottom-up estimation* – Use the detailed quantity take-off to calculate costs for individual components of the project and then aggregate them to determine the overall project cost. This bottom-up approach allows for a more accurate assessment of costs.
- *Parametric cost estimation* – Continue using parametric cost estimation techniques but refine the models with more detailed project-specific data. These models use historical data and regression analysis to estimate costs based on project parameters.
- *Comparative bidding* – For major equipment and infrastructure items, obtain competitive bids from multiple vendors. This approach helps ensure that the cost estimates are based on current market rates.
- *Engineering estimates* – Engage engineering consultants with expertise in mining to develop detailed cost estimates for various project aspects, including mine planning, processing facilities, infrastructure and environmental management.
- *Cost capacity curves* – Use cost capacity curves, which depict the relationship between production capacity and unit costs, to estimate costs at different levels of production. This helps assess the effect of scale on costs.
- *Contingency allowances* – Include contingency allowances in the cost estimates to account for uncertainties and unforeseen events during the project development. The level of contingency should be based on the project's risk assessment.
- *Risk analysis* – Perform risk analysis on the cost estimates to quantify the potential range of costs under different risk scenarios. This helps stakeholders understand the project's sensitivity to various factors.
- *Value engineering* – Consider value engineering practices to optimise costs without compromising the project's technical integrity. This involves evaluating alternatives to find the most cost-effective solutions.
- *Currency and inflation adjustment* – If the project is in a different country or currency, adjust the cost estimates for currency exchange rates and inflation factors to ensure consistency.

It is essential to maintain transparency in the cost estimation process and document all assumptions and methodologies used. Regularly review and update cost estimates as the project progresses through the Pre-Feasibility Study to ensure accuracy and reflect any changes in project scope or market conditions.

Feasibility Studies

Feasibility Studies require the most accurate and detailed cost estimation techniques. These studies involve a comprehensive evaluation of the technical, economic, social and environmental aspects of a mining project before making a final investment decision. Here are some advanced cost estimation techniques commonly used in mining Feasibility Studies:

- *Engineering, procurement and construction or construction management quotes* – Obtain detailed quotes from engineering, procurement and construction companies that specialise in design and construction of mining / Mineral Resources projects. These quotes provide accurate cost data based on the specific requirements and scope of the project. It is noted that additional engineering may be required following the Feasibility phase in order to obtain these quotes.
- *Detailed quantity take-off and bill of quantities (BOQ)* – Conduct a meticulous quantity take-off to determine the precise quantities of materials, equipment and labour required for each aspect of the project. The BOQ helps itemise the costs for individual components, aiding in accurate estimation.
- *Design-based cost estimation* – Utilise detailed engineering designs and specifications to estimate costs at a granular level. This approach ensures that the cost estimates are based on the actual requirements of the project.
- *Unit cost estimation* – Calculate costs on a per-unit basis, such as cost per tonne of ore mined or cost per tonne of ore processed. This approach allows for better cost control and comparison with industry benchmarks.
- *Benchmarking* – Compare the cost estimates with similar mining projects that have been completed recently to validate the reasonableness of the figures.
- *Parametric cost estimation with advanced models* – Continue using parametric cost estimation but refine the models with project-specific data and advanced regression analysis. These models consider multiple variables that affect costs to improve accuracy.
- *Resource estimation techniques* – For projects involving Mineral Resources, use resource estimation techniques, such as geostatistical methods, to estimate the extent and grade of the mineral deposit accurately.
- *Total cost of ownership* – Consider the total cost of ownership, which includes not only initial capital expenditures but also ongoing operational and maintenance costs over the life of the mine.
- *Contingency management* – Include appropriate contingency allowances in the cost estimates to account for potential risks and uncertainties in the project.
- *Risk analysis and sensitivity analysis* – Perform rigorous risk analysis to quantify the potential effects of various risks on project costs. Sensitivity analysis helps assess the project's sensitivity to changes in critical variables.
- *Inflation and escalation factors* – Account for inflation and escalation factors to ensure that cost estimates are expressed in real terms over the project's duration.
- *Life cycle cost analysis* – Consider the full life cycle of the mining project, including closure and reclamation costs, to determine the project's overall financial feasibility.
- *Professional expertise* – Engage experienced mining professionals and consultants with domain expertise to ensure that all relevant factors are adequately considered in the cost estimation process.

Remember that mining Feasibility Studies are critical in making investment decisions, and cost estimation accuracy is vital for project success. The methods used should be well-documented, transparent and regularly reviewed and updated to reflect any changes in project scope or market conditions.

CAPITAL AND OPERATING COST EVALUATION

Capital and operating cost estimation is a crucial aspect of Scoping, Pre-Feasibility and Feasibility Studies. It involves assessing the initial capital investment required for setting up the mine and ongoing sustaining capital and operating costs throughout the project's life cycle.

Basis of estimate

The work completed in developing the studies at each phase of project development, the Scoping Study, Pre-Feasibility Study and Feasibility Study, forms the basis of estimates of time, effort and resources required to achieve the objectives of the project. The preceding chapters describe this work.

Cost estimation quantifies the expenditure required to achieve the planned profile of effort, materials and services consumption over the life of the project.

As the project develops through the study phases, the underlying basis of the estimates improves in definition and confidence.

Cost estimates should strive to be completed at each phase of study, with no gaps in the estimate. At early phases of study, if an aspect is not well defined, an appropriate estimate should be made, rather than neglecting to consider that element.

The quality of the work used as the basis of the cost estimate directly affects the quality of the cost estimate itself. When cost estimates are reviewed as part of quality assurance, the confidence in the underlying work should be included in that review.

Capital cost evaluation

Capital costs are typically identified as fixed, one-time expenses incurred on the establishment and construction of a project to bring it to a commercial operating status. Precise definitions of capital expenditure are specific to company policy, and jurisdictional and regulatory requirements.

Work breakdown structure

The AusIMM Monograph 27 – *Cost Estimation Handbook* describes the role of a work breakdown structure (WBS) and the development of a capital cost estimate in detail. A summary of that material is presented below.

The project scope statement included in the basis of estimate must present the WBS at least down to and including Level 3 so that it includes areas, subareas and systems. The WBS ensures that the capital and operating cost estimates address each element of the project's scope of work and the operation's basis and plan.

The capital cost estimate must be based on the project's WBS, which should follow a structure similar to:

- Level 1 – prime areas
- Level 2 – subarea of the project
- Level 3 – systems or work packages
- Level 4 – assets such as equipment or elements of construction
- Level 5 – commodity or trade-discipline components of costs
- Level 6 – type of cost (eg supply, freight, install).

Schedule

The project schedule is a necessary part of the basis of estimate. Most projects are highly sensitive to the cost–time relationship and the point of transition from capital to operations.

As the capital cost estimate depends on time, it is essential that the project schedule be aligned to the same project WBS as used for the cost estimate.

Capital cost estimate

The cost estimate for a project should include the following elements. The definition of the contents of each element will increase at each phase of study:

- costs by summary, area or departments
- costs by subareas
- costs by systems.

When the basis of the estimate is sufficiently defined, the estimate should include:

- individual equipment numbers or elements of construction work
- direct costs of equipment and elements of construction
- indirect costs clearly identified by subareas and work packages
- owner's costs clearly identified by subareas, functions and positions.

When the study gets to Feasibility, the cost estimate should include individual items for:

- construction equipment
- currency of each component of the estimated item
- direct installation and construction cost
- supply cost of the material or equipment
- freight cost to site of the material or equipment
- growth allowances
- indirect costs of labour.

In earlier phases of study, a cost allowance for the above elements should be included without the definition and based on benchmark or comparable work packages.

The presentation of the cost estimate workings should include:

- quantity and units of measure
- WBS code and estimate item number unique description
- work package number for supply and/or installation
- trade type (eg civil, mechanical)
- indirect costs separately identifiable, contractor(s) indirect cost or project indirect cost
- contingency and other provisions separately identified
- method of calculation fully described for each element
- sources of cost data fully described
- estimate split by currency for exchange rate purposes
- construction cash flow forecast
- benchmark comparisons.

Closure and rehabilitation costs

It has been described in earlier sections that the impacts on external stakeholders, the environment and society, play a significant role in the value proposition of a minerals project.

No cost estimate for a minerals project can be considered complete until the costs and effort associated with impact mitigation, rehabilitation, closure and social impact are included in the financial analysis of the project.

At the Scoping, Pre-Feasibility and Feasibility Study phases of a minerals project life, the mine closure aspects are naturally less defined than the initial elements because of the impact of time until closure is expected. Throughout the life of the project, from Scoping Study, through establishment and operations, the mine closure plan should be reviewed and updated to reflect the most likely scenario at closure.

The closure cost estimates should at least consider the following elements:

- workforce termination/reassignment/retraining
- progressive and final rehabilitation of disturbed land to be fit for an agreed post-mining land use
- deconstruction and decontamination of infrastructure
- stabilising excavations and landforms
- ongoing monitoring and management of the site until appropriate for release/handover
- social programs preparing for an agreed outcome by the end of the productive life of the project.

Each project will have its own stakeholders. These stakeholders are a key part of the process to reach agreement on the above elements on closure.

Operating cost evaluation

The operating costs for a mining project can be broadly categorised into three main areas: mining costs, processing costs, and general and administrative costs. The following are some of the key factors that are typically considered in estimating the operating costs for a study:

1. *Mining costs* – This includes costs associated with mine development, drilling, blasting, excavation, hauling and reclamation. The key factors that affect mining costs include the type of mining method, the depth and geometry of the deposit, the type and size of equipment required, and the labour and energy costs.
2. *Processing costs* – This includes costs associated with processing the ore, such as crushing, grinding, flotation and leaching. The key factors that affect processing costs include the type and size of processing equipment required, the complexity of the processing flow sheet, and the energy and reagent costs.
3. *General and administrative costs* – This includes costs associated with maintaining the mine site, including salaries, wages and benefits for personnel, as well as overhead costs such as insurance, taxes and permits.

General and administrative (G&A) costs are an important component of cost estimation and are not typically detailed in the more technically focused components of the study process. G&A costs include the expenses associated with the general management, administration and support functions of a mining company that are not directly related to production or operations. Below are key considerations for evaluating G&A costs in cost estimation:

- personnel
- office and facilities
- professional services
- information systems and technology
- travel and entertainment

- corporate governance and compliance
- insurance and risk management
- investor relations and marketing
- training and development
- miscellaneous administrative expenses.

It is important to review historical data and industry benchmarks, and consider the specific organisational structure and requirements of the mining company to accurately estimate G&A costs. Regular monitoring, analysis and adjustment of G&A expenses are necessary to control costs and ensure the efficient operation of the administrative functions within the mining company.

Data sources

Cost estimation relies on various data sources to ensure accuracy and reliability. Here are some common data sources used in cost estimation:

- *Internal data* – Mining companies collect and maintain internal data that can be valuable for cost estimation. This includes historical cost data from previous projects or operations, actual expenditure records, financial statements and internal reports. Internal data provides insights into the company's specific cost structure, performance trends and resource utilisation.
- *Industry reports and publications* – Many reputable organisations and industry associations publish reports, studies and cost benchmarks specific to the mining industry. These reports provide aggregated data and key performance indicators (KPIs) related to mining costs. Examples include reports from mining associations, consulting firms, government agencies and financial institutions.
- *Engineering studies and feasibility reports* – Engineering studies and feasibility reports conducted for mining projects often contain detailed cost information. These studies outline the required infrastructure, equipment, labour and other resources, along with associated costs. They provide a basis for estimating capital expenditures (CAPEX) and operating expenditures (OPEX) for similar projects.
- *Supplier and contractor quotations* – Mining companies often engage with suppliers and contractors for various goods and services. Requesting quotations from multiple suppliers and contractors can provide cost information related to equipment, materials, consumables, services and specialised mining activities. These quotations help in estimating specific cost components accurately.
- *Cost databases and software tools* – There are specialised databases and software tools available that provide access to comprehensive cost data specific to the mining industry. These tools compile cost information from multiple sources and provide standardised cost models, estimating templates and benchmarking data. Examples include mining cost databases, cost estimation software and financial modelling tools.
- *Government and regulatory agencies* – Government agencies and regulatory bodies may require mining companies to report financial and cost information as part of compliance obligations. Publicly available filings, such as annual reports, financial statements and regulatory disclosures, can provide valuable data for cost estimation.
- *Market and commodity data* – Mining costs can be influenced by market conditions and commodity prices. Monitoring and analysing market data, such as metal or mineral

prices, fuel prices, currency exchange rates and other relevant indices, can help estimate cost variations and potential impacts on the mining operation.

- *Industry experts and consultants* – Engaging industry experts, consultants or specialised mining engineering firms can provide valuable insights and expertise in cost estimation. These professionals bring experience from similar projects, industry knowledge and access to additional data sources that can enhance the accuracy of cost estimates.

When utilising data sources for cost estimation, it is crucial to ensure the reliability, currency and relevance of the data. Data validation, cross-referencing and data quality checks should be performed to minimise errors and ensure the integrity of the cost estimation process.

Operational readiness

Operational readiness is a crucial factor in cost estimation in Feasibility Studies. It refers to the state of preparedness of a mining project or operation to transition from the construction or development phase to full-scale production. Operational readiness activities aim to ensure that all necessary systems, processes, infrastructure and personnel are in place for a smooth and efficient start of operations. Consider the following key aspects of operational readiness in cost estimation:

- *Infrastructure readiness* – Evaluate the readiness of essential infrastructure elements, such as access roads, power supply, water management systems, communication networks and waste management facilities. Estimate the costs associated with constructing, upgrading or modifying the infrastructure to meet operational requirements.
- *Equipment and asset readiness* – Assess the readiness of mining equipment, machinery and other assets required for production. This includes equipment commissioning, testing and maintenance activities. Estimate costs associated with equipment procurement, installation, training and ongoing maintenance.
- *Workforce and human resources* – Ensure that the necessary workforce is in place and adequately trained to support operations. Estimate costs associated with recruitment, training, competency assessment, and any additional staffing requirements during the operational phase.
- *Health, safety and environmental readiness* – Evaluate the readiness of health, safety and environmental management systems and procedures. Estimate costs associated with implementing safety protocols, conducting risk assessments, providing personal protective equipment and complying with environmental regulations.
- *Operational systems and processes* – Assess the readiness of operational systems and processes, including production planning, scheduling, reporting, inventory management, maintenance management and quality control. Estimate costs associated with implementing or optimising these systems, including software, hardware and training.
- *Supply chain and logistics* – Evaluate the readiness of supply chain and logistics functions to support production. This includes estimating costs associated with procuring raw materials, establishing supplier relationships, developing transportation and warehousing capabilities, and optimising inventory management.
- *Stakeholder engagement and community relations* – Consider the readiness of stakeholder engagement plans and community relations strategies. Estimate costs associated with community development programs, social and environmental impact assessments, and ongoing engagement initiatives.

- *Risk management and contingency planning* – Assess the readiness of risk management processes and contingency plans. Estimate costs associated with implementing risk mitigation measures, emergency response systems and insurance coverage.
- *Commissioning and start-up activities* – Consider the costs associated with commissioning and start-up activities, including equipment testing, performance trials and process optimisation. Estimate costs associated with initial production ramp-up, performance monitoring and fine-tuning of operations.
- *Monitoring and evaluation* – Develop mechanisms to monitor and evaluate operational readiness progress. Regularly assess the readiness of each aspect and adjust cost estimates accordingly. This includes conducting readiness audits, benchmarking against industry standards and incorporating lessons learned from similar mining projects.

Operational readiness is essential to minimise start-up delays, unexpected costs and production disruptions. Early identification and mitigation of potential operational issues during the cost estimation phase can lead to more accurate cost estimates and smoother project execution. It is important to involve cross-functional teams, including operations, engineering, maintenance and project management, in the assessment and planning for operational readiness.

Sensitivity

Mining cost estimation involves a detailed analysis of various factors and variables. A sensitivity analysis is a tool used in mining cost estimation to evaluate how changes in certain input variables affect the overall cost of the project. A sensitivity analysis helps to identify the key drivers of project costs and assess the potential risks and uncertainties associated with the project.

The following are some of the key input variables that are typically considered in mining cost estimation and sensitivity analysis:

- *Mineral Resource estimation* – The quantity and quality of the Mineral Resource are the key drivers of mining costs. A sensitivity analysis is conducted to assess how changes in the Mineral Resource estimate affect the overall project cost.
- *Mining method* – The choice of mining method, such as open pit or underground mining, will have a significant effect on the overall cost of the project. A sensitivity analysis is conducted to assess how changes in the mining method affect the project cost.
- *Production rate* – The production rate, or the amount of ore processed per unit of time, is another key driver of mining costs. A sensitivity analysis is conducted to assess how changes in the production rate affect the project cost.
- *Equipment selection* – The type and size of equipment required for mining and processing are important drivers of project costs. A sensitivity analysis is conducted to assess how changes in equipment selection affect the project.
- *Operating costs* – The operating costs associated with mining and processing, including labour, energy and materials costs, are important drivers of project costs. A sensitivity analysis is conducted to assess how changes in operating costs affect the project cost and the eventual economic viability of the project.
- *Capital costs* – The capital costs associated with building and commissioning the mine, including infrastructure, equipment and construction costs, are another key driver of project costs. A sensitivity analysis is conducted to assess how changes in capital costs affect the project cost and the eventual economic viability of the project.

By conducting a sensitivity analysis on these key input variables, the project team can identify the factors that have the most significant effect on project costs and assess the potential risks and uncertainties associated with the project. This information can then be used to optimise the project design and mitigate potential risks and uncertainties.

ACCURACY AND CONTINGENCY

Capital and operating cost estimation can be a complex task, and the accuracy of cost estimates improves as the project progresses from the Scoping phase to the Pre-Feasibility and Feasibility phases. Contingency planning is also an essential aspect of cost estimation to account for uncertainties and potential risks.

It is important to note that accuracy and contingency allowances can vary depending on the specific characteristics of the mining project, its location, complexity, and other factors. Cost estimation is subject to various risks and uncertainties, such as commodity price volatility, regulatory changes, unforeseen technical challenges and market conditions. It is crucial to regularly review and update cost estimates throughout the project life cycle to ensure that they reflect the most current information and address any emerging risks.

Accuracy

The accuracy of mineral project cost estimation depends on the quality and completeness of the data used in the estimation process. The accuracy of cost estimation can vary depending on several factors, including:

- *Resource estimation accuracy* – The accuracy of the Mineral Resource estimation is critical in determining the accuracy of the cost estimation. The more accurate the resource estimation, the more accurate the cost estimation will be.
- *Project phase* – The accuracy of the cost estimation can also vary depending on the phase of the project. At the early phases of the project, the cost estimation may be less accurate due to the lack of detailed information about the Mineral Resource, mining method, equipment selection, processing method and other key input variables. As the project progresses, the accuracy of the cost estimation typically improves.
- *Data quality* – The quality and completeness of the data used in the cost estimation process can also affect the accuracy of the estimate. Accurate and complete data about the Mineral Resource, mining method, equipment selection, operating costs and capital costs are essential to producing an accurate cost estimation.
- *Assumptions and uncertainties* – The cost estimation process involves making several assumptions about the project, including the Mineral Resource grade, the production rate, mining and processing methods, and the operating and capital costs. These assumptions can introduce uncertainties into the cost estimation, which can affect its accuracy.

To improve the accuracy of cost estimation, it is important to use high-quality data and to conduct sensitivity analyses to evaluate the effect of changes in key input variables on the cost estimate. It is also important to use experienced and knowledgeable professionals who have a deep understanding of mining engineering, geology, construction, operations and project management to conduct the cost estimation process. Finally, it is important to continually monitor and update the cost estimation as the project progresses to ensure its accuracy.

Scoping Studies

Scoping Studies are the initial phase of project evaluation. They are typically based on limited data and information and are meant to provide a rough indication of the project's economic potential. Managing accuracy in mining Scoping Studies is essential to provide a reasonable and reliable assessment of a project's potential without investing excessive time and resources. While Scoping Studies are preliminary in nature and involve limited data, there are several strategies to improve accuracy within the constraints of this early phase. The cost estimates at this phase are very high level and are often based on analogies to similar projects. As a result, the accuracy of cost estimates in Scoping Studies is generally low, with a wide range of uncertainty. Cost estimates during the conceptual study are typically presented as rough order-of-magnitude figures. They provide a broad understanding of the potential costs involved in the project but are subject to significant uncertainties. The accuracy of cost estimates at this phase can range widely, usually within a range of ±25 to 50 per cent. By managing accuracy effectively in Scoping Studies, stakeholders can gain a better understanding of the project's potential, identify potential risks, and make more informed decisions about the project's advancement to the next phase of evaluation. As the project progresses, data collection efforts, detailed engineering and more rigorous cost estimation techniques in subsequent Feasibility Studies will lead to even greater accuracy and confidence in the project's economic viability.

Pre-Feasibility Study

Managing accuracy in mining Pre-Feasibility Studies is crucial as this phase involves more detailed data and analysis compared to Scoping Studies. The goal is to provide a more refined assessment of the project's economic viability while acknowledging the remaining uncertainties. At the Pre-Feasibility phase, more data is collected through drilling programs, geotechnical studies and metallurgical testing. This leads to more accurate estimates of resource quantity and quality. The mine design becomes more detailed, allowing for better cost estimation. As a result, the accuracy of cost estimates increases, and the uncertainty range narrows. As the project advances to the Pre-Feasibility phase, more detailed engineering and analysis are conducted, leading to increased accuracy in cost estimation. Cost estimates at this phase typically have an accuracy range of ±15 to 25 per cent. Managing accuracy effectively in Pre-Feasibility Studies provides stakeholders with a clearer understanding of the project's economic potential and allows for informed decision-making regarding advancing the project, securing funding and executing the project successfully.

Feasibility Study

Managing accuracy in mining Feasibility Studies is crucial to ensure that the projected outcomes align with the actual results of the mining project. The Feasibility Study is the most comprehensive and detailed evaluation of a mining project. By this phase, a significant amount of data has been gathered, and the project's technical and economic viability is thoroughly assessed. The mine design is finalised, and detailed engineering plans are developed. Cost estimates are based on firm quotes and bids from suppliers and contractors, resulting in the highest level of accuracy and the least amount of uncertainty. The Feasibility Study represents the most detailed and accurate phase of cost estimation. Extensive engineering design, procurement planning and an operational analysis are conducted, leading to refined cost estimates. The accuracy of cost estimates at this phase is typically within a range of ±10 to 15 per cent. It is essential to recognise that even at the Feasibility phase, there will always be some level of uncertainty due to unforeseen factors that may arise during project development. Thus, it is crucial to regularly update and refine cost estimates as the project progresses, incorporating new data and addressing potential changes in market conditions, regulations or other variables. The accuracy of cost estimates directly influences the confidence in a project's economic viability and plays a pivotal role in securing funding and making informed investment decisions.

The lowest confidence level of an estimate achieved for any area should decide the overall confidence achieved for the estimate. Should one or more individual cost items not achieve the stated accuracy levels, but other items exceed the stated accuracy level, then the overall confidence in that area may be achieved for the whole project estimate.

Contingency

Mining cost estimation contingency refers to an additional amount of funds added to the estimated mining cost to account for unforeseen risks and uncertainties that may affect the project's cost. Contingency funds are typically added to the project cost estimate to provide a buffer against unexpected cost overruns and to ensure that the project remains financially viable.

The amount of contingency added to the cost estimate will depend on the level of risk and uncertainty associated with the project. Generally, the contingency amount is a percentage of the total estimated cost and can vary from project to project.

It is important to note that the contingency amount is not a fixed number and can change over the course of the project. As the project progresses and more information becomes available, the contingency amount can be adjusted to reflect the changing risk profile of the project.

The contingency amount can be used to cover a range of known unforeseen costs, including changes in the Mineral Resource grade, unexpected geological conditions, equipment breakdowns and changes in regulatory requirements. By including a contingency amount in the cost estimate, project managers can ensure that the project remains financially viable and that unexpected cost overruns do not derail the project.

The recommended amount of contingency is described at each study phase in Table 9.1. If a different amount is presented, a sufficiently detailed justification should accompany the estimate.

Overall, adding a contingency amount to the cost estimate is an important risk management tool that can help ensure the financial success of the project.

Scoping Studies

Managing contingency in Scoping Studies requires a different approach compared to Feasibility Studies due to the limited data and preliminary nature of the Scoping phase. While Scoping Studies involve a more conceptual assessment of a mining project's potential, it is still essential to address uncertainties and potential risks. Due to the limited amount of detailed information available during the conceptual stage, it is important to include contingency allowances in the cost estimates. Individual contingency percentages can range from 20 to 30 per cent to account for potential variations and unknowns. The primary goal of a Scoping Study is to assess the potential of the mining project and determine if it is worth advancing to more detailed phases of study. While the level of accuracy and quantification of contingency is limited at this phase, managing uncertainties and risks in

Scoping Studies is a proactive approach that can lead to more informed decision-making and better-prepared stakeholders as the project evolves.

Pre-Feasibility Studies

Managing contingency in Pre-Feasibility Studies is an essential step in ensuring a more accurate assessment of the project's economic viability. At the Pre-Feasibility phase, more data and information are available compared to Scoping Studies, allowing for a more detailed evaluation of costs and risks. Contingency allowances are still necessary at the Pre-Feasibility phase, although they may be reduced compared to the conceptual stage. Individual contingency percentages can range from 15 to 30 per cent, considering remaining uncertainties and risks. Managing contingency effectively in mining Pre-Feasibility Studies helps to provide a more realistic financial outlook for the project. It demonstrates a proactive approach to risk management and financial planning, which is crucial for gaining stakeholder confidence and attracting potential investors or financing partners. As the project progresses to the Feasibility phase, the contingency estimates should be further refined based on more detailed information and data.

Feasibility Studies

Managing contingency is a critical aspect of Feasibility Studies as it helps account for uncertainties and risks that could affect the project's cost estimates. Individual contingency allowances in the Feasibility Study are generally lower compared to earlier phases. They typically range from 10 to 15 per cent and account for any remaining uncertainties, unexpected events or cost overruns that may occur during the project's implementation. Managing contingency effectively in Feasibility Studies provides a safety net against unforeseen circumstances and helps stakeholders gain confidence in the project's financial viability. It also demonstrates a proactive approach to risk management and financial planning, which is essential for securing funding and ensuring successful project execution.

Summary

Scoping Study

- *Accuracy* – Cost estimates during the conceptual study are typically presented as rough order-of-magnitude figures. They provide a broad understanding of the potential costs involved in the project but are subject to significant uncertainties. The accuracy of cost estimates at this phase can range widely, usually within a range of ±25 to 50 per cent.
- *Contingency* – Due to the limited amount of detailed information available during the conceptual stage, it is important to include contingency allowances in the cost estimates. Contingency percentages can range from 25 to 30 per cent to account for potential variations and unknowns.

Pre-Feasibility Study

- *Accuracy* – As the project advances to the Pre-Feasibility phase, more detailed engineering and analysis are conducted, leading to increased accuracy in cost estimation. Cost estimates at this phase typically have an accuracy range of ±15 to 25 per cent.
- *Contingency* – Contingency allowances are still necessary at the Pre-Feasibility phase, although they may be reduced compared to the conceptual stage. Contingency percentages can range from 15 to 30 per cent, considering remaining uncertainties and risks.

Feasibility Study

- *Accuracy* – The Feasibility Study represents the most detailed and accurate stage of cost estimation. Extensive engineering design, procurement planning and an operational analysis are conducted, leading to refined cost estimates. The accuracy of cost estimates at this phase is typically within a range of ±10 to 15 per cent.
- *Contingency* – Contingency allowances in the Feasibility Study are generally lower compared to earlier phases. They typically range from five to 15 per cent and account for any remaining uncertainties, unexpected events or cost overruns that may occur during the project's implementation.

RISK MANAGEMENT IN CONTINGENCY PLANNING

Risk management and contingency planning are integral parts of cost estimation in feasibility studies. Mining projects involve various risks and uncertainties that can affect costs, schedules and overall project success. Effective risk management and contingency planning help identify potential risks, assess their potential impact, and develop strategies to mitigate or manage them. Here is how risk management and contingency planning are applied in cost estimation:

- *Risk identification* – Identify and categorise potential risks specific to the mining project or operation. These risks can include geological uncertainties, commodity price fluctuations, regulatory changes, environmental impacts, labour shortages, equipment failures or community opposition. Involve stakeholders and subject matter experts to ensure comprehensive risk identification.
- *Risk assessment* – Assess the likelihood and potential effect of identified risks on the project's objectives and cost estimates. Quantify risks using techniques such as probability analysis, sensitivity analysis or scenario analysis. Prioritise risks based on their significance and potential consequences.
- *Risk mitigation strategies* – Develop strategies and action plans to mitigate or reduce identified risks. This may involve implementing preventive measures, such as improved geological surveys, robust safety protocols or redundant equipment. Explore alternative approaches, technologies or suppliers to reduce dependence on specific risk-prone factors. Consider risk-sharing mechanisms, such as insurance or hedging strategies, to transfer certain risks.
- *Contingency planning* – Develop contingency plans to address potential risks that cannot be fully mitigated or eliminated. Contingency plans outline specific actions, resources and timelines to be activated in response to identified risks. Allocate appropriate budgets for contingencies to cover additional costs that may arise due to risk events.
- *Cost impact assessment* – Estimate the potential cost impact of identified risks on the project's budget. Quantify the potential cost overruns, delays or additional expenditures that may result from risk events. Consider both direct costs, such as repairs or replacements, and indirect costs, such as downtime, rescheduling or increased financing expenses.
- *Risk monitoring and control* – Continually monitor and reassess identified risks throughout the project life cycle. Establish monitoring mechanisms, such as regular risk assessments, KPIs or early warning systems, to identify emerging risks or changes in risk profiles. Implement control measures to manage identified risks and update cost estimates accordingly.

- *Communication and reporting* – Maintain open and transparent communication channels with stakeholders regarding identified risks, mitigation strategies and contingency plans. Provide regular risk status updates, highlighting changes, and the associated cost implications. Incorporate risk management and contingency planning discussions in project reviews or progress reports.
- *Lessons learned* – Capture lessons learned from risk events or unexpected cost deviations. Document and analyse the root causes, response effectiveness and cost impacts of risk events. Use this information to continually improve risk management processes and enhance future cost estimation accuracy.

By integrating risk management and contingency planning into cost estimation, mining companies can enhance their ability to anticipate and respond to potential risks. This proactive approach helps mitigate the effect of risks on project costs, improve project outcomes and ensure more accurate cost estimates. Risk management should be an ongoing process throughout the project life cycle, with regular reviews and updates to adapt to changing circumstances and emerging risks.

PEER REVIEW

Peer review is an essential component of the cost estimation process in Feasibility Studies. It involves obtaining an independent evaluation and validation of the cost estimates from knowledgeable professionals or experts within the field. Peer review helps to ensure the accuracy, reliability and completeness of the cost estimation, and it helps identify any potential errors, biases or oversights. Here is how peer review is typically conducted:

1. *Selection of reviewers* – The mining company or the cost estimation team selects qualified reviewers who possess relevant experience and expertise in mining cost estimation. Reviewers may include professionals from within the organisation, external consultants, industry experts or peers from other mining companies.
2. *Documentation sharing* – The cost estimation team provides the reviewers with the necessary documentation, including the cost estimation methodology, assumptions, data sources, calculations and supporting information. This documentation should be comprehensive and transparent to facilitate a thorough review.
3. *Review scope* – The scope of the peer review is defined, outlining the specific aspects of the cost estimation to be evaluated. This may include reviewing the accuracy of cost assumptions, appropriateness of cost drivers, completeness of cost components, consistency with industry benchmarks and compliance with established standards.
4. *Independent analysis* – Reviewers independently analyse the cost estimation documentation and perform their own calculations or assessments to validate the cost estimates. They may conduct sensitivity analysis, challenge assumptions, verify data sources and compare the estimates with their own knowledge and experience.
5. *Review feedback* – The reviewers provide feedback, comments and recommendations on the cost estimation. This feedback may include identifying potential errors, suggesting alternative cost drivers or methodologies, pointing out areas of uncertainty or risk, and providing insights for improvement. The feedback should be specific, constructive and supported by reasoning or evidence.
6. *Review discussion* – The cost estimation team engages in discussions with the reviewers to further clarify and address any concerns or questions raised during the review process.

This interaction allows for a deeper understanding of the review findings and facilitates the resolution of any discrepancies.

7. *Revision and finalisation* – Based on the feedback and discussions, the cost estimation team revises the cost estimates, addressing the issues raised by the reviewers. They incorporate the recommendations or modifications that enhance the accuracy and reliability of the cost estimation. The final cost estimates are then documented and ready for use in decision-making processes.

Peer review enhances the quality and credibility of cost estimation by bringing in independent perspectives and expertise. It helps identify potential biases, errors or weaknesses that may have been overlooked in the initial estimation. By incorporating the insights and recommendations from the peer review process, mining companies can improve the accuracy of their cost estimates and enhance their decision-making processes.

BENCHMARKING

Benchmarking is a valuable practice in cost estimation. It involves comparing the costs, performance and practices of a mining project or operation against industry peers or established standards. By benchmarking, mining companies can gain insights into their cost position and identify areas for improvement. Here are the key aspects of benchmarking in cost estimation:

1. *Cost categories* – Identify the specific cost categories to benchmark, such as CAPEX, OPEX, G&A and specific cost components within these categories. This could include costs related to labour, equipment, maintenance, energy, consumables and administrative expenses.
2. *Selection of peers* – Identify comparable mining projects or operations that are similar in terms of size, type of mineral deposit, mining method, geographic location and other relevant factors. These can be mining operations within the same company, competitors or industry benchmarks published by reputable sources.
3. *Data collection* – Gather data on the costs and performance metrics of the benchmarked projects or operations. This may involve accessing publicly available data, industry reports and financial statements, and engaging in industry networks or associations. Confidentiality and data sharing agreements may be required when accessing proprietary information.
4. *Data analysis* – Analyse the benchmarking data to identify the gaps and variations between the mining company's costs and those of the benchmarked projects or operations. Compare key cost metrics, such as cost per unit of production, cost per tonne of ore processed, or cost per ounce of metal produced. Look for significant deviations and areas of opportunity or concern.
5. *Root cause analysis* – Conduct a detailed analysis to understand the factors behind the differences in costs. Identify the drivers of cost variation, such as differences in labour productivity, equipment utilisation, energy efficiency, procurement practices or economies of scale. Determine the root causes that contribute to the cost gaps.
6. *Best practices identification* – Identify best practices from the benchmarked projects or operations that could be adopted or adapted to improve cost performance. This may involve analysing the approaches, strategies, technologies and processes that contribute to lower costs or higher efficiencies in the benchmarked cases. Determine the feasibility and potential effect of adopting these practices in the mining company's operations.

7 *Performance improvement planning* – Develop an action plan based on the benchmarking analysis and best practices identified. Prioritise improvement opportunities and establish specific targets and initiatives to address the identified cost gaps. Engage relevant stakeholders, including operations, finance, engineering and management teams, to drive the implementation of improvement initiatives.
8. *Regular monitoring and review* – Continually monitor and review the cost performance of the mining company against the benchmarks. Track the progress of improvement initiatives and adjust strategies as needed. Regularly update the benchmarking analysis to ensure ongoing alignment with industry standards and changes in market conditions.

Benchmarking provides valuable insights for mining companies to understand their cost competitiveness and identify opportunities for cost reduction or operational improvement. However, it is important to consider the specific context and unique characteristics of each mining operation when interpreting benchmarking results and applying them to cost estimation.

AUDITING

Auditing plays a critical role in evaluating and verifying the accuracy of cost estimates. Auditing helps ensure that mining companies adhere to sound financial practices and provides assurance to stakeholders that cost estimates are reliable. Here are key aspects of cost estimation auditing:

1. *Independent audit* – Auditing is typically conducted by an independent third party, such as a professional auditing firm. This ensures objectivity and impartiality in the audit process.
2. *Documentation review* – Auditors carefully review the documentation related to cost estimation, including Feasibility Studies, engineering reports, financial models, contracts and supporting data. They assess the completeness, accuracy and reliability of the information used in the cost estimation process.
3. *Assumptions and methodologies* – Auditors scrutinise the assumptions and methodologies employed in cost estimation. They evaluate the reasonableness of key assumptions, such as commodity prices, exchange rates, operating costs and discount rates. The auditors verify that the selected methodologies align with industry standards and best practices.
4. *Comparisons and benchmarking* – Auditors may compare the cost estimates with similar mining operations, industry benchmarks and historical data to identify any significant discrepancies or outliers. This helps assess the reasonableness of the estimated costs and highlights areas for further investigation.
5. *Cost categories and allocation* – Auditors examine the breakdown of costs into various categories, such as CAPEX and OPEX. They ensure that costs are allocated appropriately and in accordance with accepted accounting principles.
6. *Verification of supporting data* – Auditors validate the accuracy and reliability of supporting data used in the cost estimation process. This includes reviewing invoices, contracts, purchase orders, vendor agreements and other relevant documents to verify the existence and validity of costs incurred.
7. *Internal controls and processes* – Auditors assess the adequacy of internal controls and processes related to cost estimation. They review the mining company's systems for cost tracking, reporting and control, ensuring that there are proper mechanisms in place to prevent fraud, errors or misstatements.

8. *Reporting and recommendations* – After completing the audit, auditors prepare a report that outlines their findings and recommendations. The report may include observations on the accuracy and reliability of the cost estimates, any identified deficiencies or weaknesses in the estimation process, and suggestions for improvement.

Auditing cost estimates promotes transparency, enhances credibility and helps in risk management. It provides stakeholders, including investors, lenders and regulatory authorities, with confidence in the financial viability and integrity of mining projects.

EXECUTION PLAN

An execution plan is a crucial component of cost estimation for an advanced project. It outlines the specific activities, timelines, resources and budgets required to bring the mining project into operation and achieve the estimated costs. Here are key considerations in developing an execution plan for cost estimation:

1. *Project phases* – Divide the project into distinct phases, such as exploration, development, construction and production. Each phase will have its own set of activities, deliverables and cost components. Define the objectives and milestones for each phase.
2. *Work breakdown structure* – Create a comprehensive WBS that breaks down the project into smaller, manageable tasks. Each task should be clearly defined, with specific deliverables, duration, dependencies and responsible parties. The WBS serves as a foundation for estimating costs and resource requirements.
3. *Resource allocation* – Identify the resources required for each task, such as labour, equipment, materials and services. Determine the quantity and availability of these resources, considering factors like skill sets, equipment availability and procurement lead times. Allocate resources based on the project schedule and estimated costs.
4. *Schedule development* – Develop a detailed schedule that specifies the start and end dates for each task, as well as the interdependencies between tasks. The schedule should consider factors like seasonal variations, regulatory approvals and any external constraints. Align the schedule with the estimated costs to ensure timely execution.
5. *Cost estimation* – Based on the WBS and resource allocation, estimate the costs associated with each task and phase. Consider direct costs (eg labour, equipment, materials) as well as indirect costs (eg overheads, administration, contingencies). Utilise historical data, industry benchmarks and input from subject matter experts to ensure accurate cost estimation.
6. *Budget development* – Aggregate the estimated costs for each task and phase to develop a project budget. Include allowances for potential cost variations, inflation and unforeseen events. The budget should align with the overall financial objectives of the project and provide a clear cost baseline for tracking and control.
7. *Risk assessment and mitigation* – Identify potential risks and uncertainties that may affect the execution plan and cost estimation. Assess the likelihood and potential effect of these risks and develop mitigation strategies to minimise their effects. Regularly review and update the risk assessment throughout the project life cycle.
8. *Monitoring and control* – Implement systems and processes to monitor and control project execution and costs. Establish KPIs to measure progress and performance against the plan. Regularly track actual costs, compare them with estimated costs, and analyse any variances. Implement corrective actions as necessary to address deviations from the plan.

9. *Communication and reporting* – Establish clear communication channels and reporting mechanisms to keep stakeholders informed about the execution plan and cost performance. Provide regular updates on progress, milestones achieved, and any significant changes or deviations from the plan. Transparency in reporting builds trust and allows for timely decision-making.

The execution plan is a dynamic document that should be regularly reviewed, updated and adjusted as the project progresses. It serves as a roadmap for effective implementation, ensuring that the estimated costs are aligned with the project's objectives and are executed in a controlled and efficient manner.

CONCLUSIONS – STUDY PROCESS AS A BASIS FOR COST ESTIMATION

The cost estimation process involves assessing and forecasting the expenses associated with the exploration, development and production of Mineral Resources. It is crucial for mining companies to accurately estimate costs to make informed decisions, secure financing and evaluate project feasibility. Here is an overview of the typical steps involved in the cost estimation process:

1. *Scope definition* – Define the scope of the project, including the size and nature of the mineral deposit, the proposed mining method and the production targets. This provides the basis for cost estimation.
2. *Resource evaluation* – Conduct geological surveys and analysis to determine the quantity and quality of Mineral Resources present in the deposit. This includes estimating the ore grade, tonnage and mineral content, as these factors affect the overall mining costs.
3. *Mine planning and design* – Develop a mine plan that outlines the sequence of mining operations, infrastructure requirements, equipment selection and mine life. The design considers factors such as geotechnical considerations, ore characteristics and environmental regulations.
4. *Capital expenditure estimation* – Determine the initial investment required to establish the mining operation. This includes costs associated with land acquisition, site preparation, mine infrastructure, equipment purchases and construction. Detailed engineering studies and cost models are often utilised for accurate CAPEX estimation.
5. *Operating expenditure estimation* – Assess the ongoing costs of mining operations, including labour, energy, consumables, maintenance and repairs. OPEX estimation also involves evaluating the costs associated with processing, transportation, waste management and environmental compliance.
6. *Financial analysis* – Perform financial evaluations, including net present value (NPV), internal rate of return (IRR) and payback period calculations, to assess the economic viability of the project. Sensitivity analysis may be conducted to understand the effect of changing cost assumptions on project economics.
7. *Cost monitoring and control* – Implement systems and processes to monitor and control costs during the different stages of mining. This includes regular tracking and reporting of actual costs against budgeted costs, identifying cost variances and implementing corrective actions as necessary.
8. *Risk assessment* – Identify potential risks and uncertainties that can affect cost estimates, such as commodity price volatility, regulatory changes, labour disputes and environmental factors. Incorporate risk management strategies and contingencies to mitigate these risks.

It is important to note that the cost estimation process may vary depending on the type of mining operation (eg underground, open pit, or placer mining) and the specific characteristics of the mineral deposit. Additionally, engaging experienced mining engineers, geologists and financial analysts is crucial to ensure accurate and reliable cost estimates.

ECONOMIC EVALUATION

10

INTRODUCTION

The process of economic evaluation is critical in the life cycle of a resources asset as it is a process that determines the expected economic viability of a project and is used as the basis for investment decisions. The use of Feasibility Studies can vary from internal decision-making at the executive or Board of Directors level or externally as a component of due diligence for project financing.

This chapter provides guidance on best practice in economic evaluation of mining projects through different phases of development from Scoping Study through to Feasibility Study. This chapter is prepared in a broad context and is not specific to any particular commodity or market; however, it aims to provide the reader with some key concepts to consider when preparing a mining project study. It also includes guidance for best-practice modelling and disclosure, which is based on our experience in advising on mining asset and company transactions as well as our interactions with industry bodies, the corporate regulators in Australia and project financiers. The guidance is also based on the level of information that is typically expected from a due diligence perspective when evaluating a mining project.

The concept of economic evaluation is applied at each study phase (refer to Figure 1.2 in Chapter 1). Where the process or requirements of economic evaluation differ between the different study phases, these are identified separately. The requirements of a Scoping Study, Pre-Feasibility Study and Feasibility Study are presented in the context of economic evaluation in the section 'Understanding the confidence level of the study', but largely the principles and recommendations provided in this chapter apply across all three study phases, except where stated otherwise.

Set out in Table 10.1 is a summary of the differences between the level of confidence for each of a Scoping Study, Pre-Feasibility Study and Feasibility Study.

STUDY CHARTER APPLICATION

The study charter is an important component to the life cycle of a mining asset as it assists with the messaging of the business case within the organisation. The messaging of the business case also applies to external stakeholders. The outcome of the study will ultimately be disclosed to the market and so should be consistent with prior messaging to the market regarding the company's strategy and direction.

TABLE 10.1

Economic evaluation study phase guide.

ECONOMIC EVALUATION					
Activity	**Deliverable**	**Description**	**SS**	**PFS**	**FS**
Preparing a study charter and messaging the business case	Understanding what a study charter is and its purpose	The relationship between the company's goals, objectives and strategy and its studies. Understanding the key elements to a study charter. Understanding the resources required to progress through the study phases.	✓	✓	✓
	Messaging the business case	Understanding the different stakeholders and communicating the business case accordingly.	✓	✓	✓
Expressing the output from the preceding chapters in the form of a financial model	Preparing a financial model	Preparing a three-way financial model that is: • flexible, adaptable and easily understood • underpinned by a schedule of supportable inputs • is free from material error – peer or external reviews conducted.	✓	✓	✓
Understanding the confidence level of the study	Ensure that the study is prepared to an appropriate level of accuracy and its content is in accordance with stated objectives	Scoping Study to be within ±25 to 50% level of accuracy. Pre-Feasibility Study to be within ±15 to 25% level of accuracy. Feasibility Study to be within ±10 to 15% level of accuracy.	✓	✓	✓
Selecting and understanding project hurdles and metrics	Evaluate the project using appropriate metrics	Consider metrics such as: • net present value • internal rate of return • payback period • cost metrics.	✓	✓	✓
Application of external factors	Impact of external factors on the viability and risks of the project	Consider the impact of the following factors on the viability of the project: • commodity prices • exchange rates • taxation • environmental factors and permitting.	✓	✓	✓
Selling the commodity and other market forces	Assess the market in which the commodity will be sold	Understand who the customer will be and the economics for the project, as early in the process as the study charter. This is to be considered with additional rigour as the project progresses through the Scoping Study, Pre-Feasibility Study and Feasibility Study phases.	✓	✓	✓
Sensitivity analysis and scenario analysis	Sensitivity analysis	Present a sensitivity analysis to show how sensitive the net present value of a project is to changes in key inputs and assumptions.	✓	✓	✓
	Scenario analysis	Present different scenarios that may be considered. This is most appropriate at the Pre-Feasibility Study phase; however, it can be used at a Scoping Study or Feasibility Study phase.		✓	
Assessment of the most appropriate and consistent presentation basis	Ensure that the model and input underpinning the model are on a consistent basis	Consider whether cash flows are presented on a nominal or real basis.	✓	✓	✓
		Determine an appropriate discount rate.	✓	✓	✓
		Consider whether project metrics are on a funded or unfunded basis.	✓	✓	✓
		Estimate the quantum of funding required as your project progresses through the study phases.	✓	✓	✓
Incorporating the requirements of project financiers	Consider funding sources	Identify potential funding sources and ensure that the model and studies are prepared accordingly. Funding is typically addressed at the Feasibility Study phase, but it should be a considered throughout the study process.	✓	✓	✓

SS = Scoping Study; PFS = Pre-Feasibility Study; FS = Feasibility Study.

Although 'spin' has its place when announcing the findings of a study to the market to highlight opportunity, it should be used carefully and within reason. Transparency is key, as an educated reader of the announcement will understand what is not included in an announcement and can differentiate spin from genuinely positive results and ensure compliance with regulatory requirements.

It is also important to understand the audience for which the study and the summary of the study is being prepared. The overview of the results of the announcement should be presented in a way that the layperson can understand. The smartest people in our industry are those who can disseminate complex and technical information in such a way that can be understood by the majority who are not technical specialists. A company will get the most benefit if the outcome of its studies can be understood by many and there is a level of transparency that users of the report can use to form their own conclusions.

EXPRESSING THE OUTPUT FROM THE PRECEDING CHAPTERS IN THE FORM OF A FINANCIAL MODEL

The financial model is considered by many (finance nerds) as the 'sexy' part of economic evaluation. Whether it be finance professionals or technical mining professionals, many get excited by a fully functional financial model. It is an imperative tool that provides the basis for a company's studies, whether it be a Scoping Study, Pre-Feasibility Study or Feasibility Study. Set out below are what are considered to be the key pillars of an effective financial model.

Three-way financial model

A three-way financial model is a model that includes a statement of profit and loss, a balance sheet and a cash flow statement. One of the characteristics of a poor life-of-mine model is that from a financial point of view it only considers cash flows. It is less important for a Scoping Study, but as a project progresses through to a Pre-Feasibility Study and Feasibility Study, it is important to understand the working capital position of the business as well as any funding shortfalls that may exist. A tool to achieve this is the balance sheet.

Simple, transparent and easily understood

The most effective financial models are those that are easily understood and where the calculations have a clear audit trail. It can often be tempting to flex your modelling muscle by incorporating complex formulae or by using macros within the model. A macro is a program within Microsoft Excel that allows an action, or a set of actions, to be performed automatically. Macros can serve a purpose but should only be used when they are required for functionality. The use of macros can often slow the model and will reduce the level of transparency of the model as the automations typically result in a 'black box' calculation.

Flexible and adaptable

Experience shows that a (less than organised) company will often build the financial model to address the level of confidence required by a Scoping Study and will not build in the ability to flex certain inputs and assumptions. As the studies progress, the builder of the original model will often tack on additional detail, which can affect the functionality of the model and can result in it going from a dynamic to a static model. This also amplifies the key person risk in that as the model becomes more cumbersome, if the person responsible for building the model leaves the organisation, there can be additional time and cost in understanding the additions to the model. This is a common source of errors when a financial model is being amended and often results in either the original model becoming cumbersome and inefficient, or the decision being made to abandon the original model and for a completely new one to be constructed.

The best-practice approach to encompassing the inputs detailed in previous chapters into the financial model is to ensure that the financial model is built to withstand the increasing requirements of the studies, from the Scoping through to Feasibility phase. It should have the functionality to be easily understood and reviewed, yet the sophistication to be able to flex key inputs and assumptions. This is detailed further in the subsection on Sensitivity Analysis and Scenario Analysis. By investing in futureproofing the model at the onset of the studies, this will save the company time and money in the long run. While it may be viewed as a potential efficiency, repurposing an existing model is often far less effective than a built-for-purpose model that commences once other aspects of the study are well understood. This ensures the model follows the physical flow of the project and starts with the user's end needs in mind in terms of output.

Further, it is common for companies to publish results of a Scoping, Pre-Feasibility or Feasibility Study using a fixed set of economic assumptions over the entire life-of-mine, for example commodity prices and exchange rates, as well as a notional discount rate to inform the net present value (NPV) of the cash flows of the project. During due diligence, a potential investor will form its own view on the inputs to the model, therefore most investors will expect that the model can be flexed and the inputs changed. If this is not the case, then the company runs the risk of limiting its investor pool, which will likely lead to a less than optimal outcome.

Supportable inputs

When building the financial model for use in a study, the inputs used for the model should be supported. The level of support will differ on a case-by-case basis, depending on the project, the input and the confidence level of the study. It is recommended that support for the inputs be organised and easily accessible. By forming these habits at the initial study phases, it reduces the risk of error. It also prepares the company and management team for due diligence, should a transaction be contemplated or external funding sought.

Experience shows that during the due diligence phase for transactions involving mining assets, a financier or potential acquirer will often require supporting information for:

- upfront and sustaining capital expenditure assumptions
- mining physicals, including volumes, grade, waste material and mine life
- recovery and processing assumptions
- offtake or other selling and marketing assumptions
- operating cost assumptions
- rehabilitation cost assumptions
- tenure of mineral assets.

Financial model that is free from error

The results of the studies in financial terms are only as good as the model (including source documents as set out above) that underpin these results. If the model that was built for the Feasibility Study is then ultimately used for sourcing funding, then it is likely that the investor will have a valuations/modelling professional review the model as part of their due diligence.

As an absolute minimum a potential financier's due diligence will test that the model provided is mathematically accurate, has structural integrity and is logical. If the model falls down in any of these areas, it is a red flag for any potential investor. It is therefore important to ensure the model's structural integrity is in order prior to engaging with potential financiers. This is detailed further in the section 'Incorporating the requirements of project financiers'.

UNDERSTANDING THE FINANCIAL IMPACT OF THE IDENTIFIED RISKS AND OPPORTUNITIES

It is important that the financial model for a mining study has the above characteristics, but it should also factor in, and be capable of flexing, the risks and opportunities of the project. The risks of the project, if well-defined and articulated in the study charter and subsequently in each phase of the mining studies, should be able to be linked to the financial model. Different users of the financial model are likely to have differing views on the key risk and opportunities.

The extent to which the risks, and importantly the costs of mitigation, are reflected in the model will depend on the level of study that is being conducted.

UNDERSTANDING THE CONFIDENCE LEVEL OF THE STUDY

When conducting an economic evaluation of a mining project, there is a broad range of commonly understood levels of accuracy for a Scoping Study, Pre-Feasibility Study and Feasibility Study, which are discussed below. These ranges represent preferred levels of accuracy; however, there may be some inputs that are unable to be reported with a sufficient level of accuracy. Transparency is key. As long as the deficiency in the level of accuracy for a particular input is clearly disclosed in the study, the user of the study can factor that in either qualitatively (incorporate in their risk assessment) or quantitatively (through an adjustment to the discount rate or adjusting the cash flows). This is also where the use of scenario analysis and sensitivity analysis is useful (refer to sections 'Sensitivity analysis' and 'Scenario analysis').

Scoping Study

A Scoping Study, as defined by the Code for Reporting of Exploration Results, Mineral Resources and Ore Reserves (the 'JORC Code', 2012), is an order of magnitude technical and economic study of the potential viability of Mineral Resources. It includes appropriate assessments of realistically assumed Modifying Factors together with any other relevant operational factors that are necessary to demonstrate at the time of reporting that progress to a Pre-Feasibility Study can be reasonably justified. A Scoping Study can also be referred to as a conceptual study, as they are exactly that, a high-level study that scopes out the project that is conceptual in nature. The purpose of the Scoping Study is to determine whether additional studies should be conducted. The confidence level of a Scoping Study is low (relative to Pre-Feasibility and Feasibility Studies), with the level of accuracy required being between ±25 and 50 per cent.

Given the nature of a Scoping Study, the investment risk is typically low, but obviously the intention is to not invest further funds into a project that has no chance of being economic. The major risk at this phase is that a viable mining project is relinquished due to an inadequate assessment. Inherently, an exploration project will have a low probability of developing into a producing asset, therefore the risk that precludes the asset from further studies would typically need to be considered a 'fatal flaw'.

There is no minimum level of geological confidence for a project in order for it to proceed to the Scoping Study phase. It will typically have a declared Mineral Resource estimate, whether it be Inferred, Indicated or Measured; however, this is not a prescriptive characteristic of a project that is entering the Scoping Study phase.

The key considerations when preparing a Scoping Study are:

- Will the project potentially provide a desirable outcome?
- Are the project and these outcomes in alignment with the company's strategy?
- Does the project help the company to meet its goals and objectives?
- If the project is economic, will the company have the ability to fund it?
- Is there sufficient upside to warrant investment in additional studies?
- What are the key risks to the project and can these potentially be mitigated?
- Are there any fatal flaws?

After considering the above, the Scoping Study should also include a workplan, detailing the process for progressing the project through to a Pre-Feasibility Study.

Pre-Feasibility Study

A Pre-Feasibility Study, as defined by the JORC Code (2012), is a comprehensive study of a range of options for the technical and economic viability of a Mineral Resource project that has advanced to a stage where a preferred mining method, in the case of underground mining, or the pit configuration, in the case of an open pit, is established and an effective method of mineral processing is determined. It includes a financial analysis based on reasonable assumptions on the Modifying Factors and the evaluation of any other relevant factors that are sufficient for a Competent Person, acting reasonably, to determine if all or part of the Mineral Resources may be converted to an Ore Reserve at the time of reporting. A Pre-Feasibility Study is at a lower confidence level than a Feasibility Study.

The purpose of a Pre-Feasibility Study is to provide a preliminary economic assessment of the various options that are considered for the project and to ultimately present a conclusion on the best option to proceed with. The level of confidence for a Pre-Feasibility Study is between ±15 and 25 per cent.

The level of disclosure and analysis included in a Pre-Feasibility Study can differ depending on the purpose of the study. A Pre-Feasibility Study can be prepared for a number of different reasons; the most common are set out below:

- as a basis for committing to a major exploration program in order to prove up a resource
- to attract a buyer, joint venture partner or investor in the project, often to fund the initial capital expenditure requirements
- to allow the company to declare a Reserve if the Modifying Factors have been addressed in compliance with the JORC Code
- in some circumstances, a Pre-Feasibility Study can be prepared by a potential acquirer as part of its due diligence process
- to compare various mining/development options and identify the most optimal and progress through to a Feasibility Study.

The main components of a Pre-Feasibility Study are:

- a mine design based on a resource model
- a discussion of the alternatives considered, and the optimal case selected
- preliminary studies completed on geotechnical, environmental and infrastructure requirements
- bench-scale metallurgical tests and preliminary process design
- cost estimates
- go/no go decision on whether the project should proceed to Feasibility Study.

In accordance with the continuous disclosure and Listing Rule obligations of companies listed on the Australian Securities Exchange (ASX), the results of a Pre-Feasibility Study will often be released to the market. Therefore, although a Pre-Feasibility Study is not expected to be a complete economic evaluation of the project, there are implications from an investor relations perspective if following the completion of the study, the project does not proceed or an alternative path to that chosen in the Pre-Feasibility Study is taken. As such, in order to reduce the risk of regulatory action or running into potential investor relations issues, it is important that companies commit the requisite resources to the completion of a Pre-Feasibility Study.

Further, companies can slip up in the disclosures to the market in the Pre-Feasibility Study phase. These studies should be performed on realistic and achievable parameters/assumptions with relevant scenarios and sensitivity analysis, showing the users of the study that there is a low confidence level with the expected outcomes, and that there is still a wide range of possible outcomes (refer to sections 'Sensitivity analysis' and 'Scenario analysis' for further information).

In messaging the market about the results of a Pre-Feasibility Study, a company should be cognisant of setting a level of expectation in the market. The market's expectation of a project can be likened to a tube of toothpaste; once the toothpaste is out of the tube, it is almost impossible to get back in. Similarly, if a company's disclosure of its results of a Pre-Feasibility Study to the market is based on assumptions that are unlikely to be achievable, it is very difficult to rein in the market's expectations of a project. The consequence of this is that if and when the company proceeds to a Feasibility Study or further through to production, the company is set out to underperform. The thing we know about capital markets is that the magnitude of decreases in value arising from an unexpected negative event or underperformance against a previously communicated target will typically be greater than the upside associated with a positive announcement or outperformance. That is, our market punishes bad news more than it rewards good news.

Feasibility Study

A Feasibility Study, as defined by the JORC Code (2012), is a comprehensive technical and economic study of the selected development option for a Mineral Resource project that includes appropriately detailed assessments of applicable Modifying Factors together with any other relevant operational factors and detailed financial analyses that are necessary to demonstrate at the time of reporting that extraction is reasonably justified (economically mineable). The results of the study may reasonably serve as the basis for a final decision by a proponent or financial institution to proceed with, or finance, the development of the project. The confidence level of the study will be higher than that of a Pre-Feasibility Study.

The Feasibility Study is usually based on the most attractive alternative for the project as previously determined and is prepared to a confidence level of ±10 to ±15 per cent. The aim of the study is to remove all significant uncertainties and to present the relevant information that can be supported, in a concise manner. The Feasibility Study has a number of key objectives:

- to demonstrate within reasonable confidence that the project can be constructed and operated economically
- to provide a basis for detailed design and construction
- to enable the raising of finance for the project from debt or equity financiers.

The Feasibility Study is one step along the design path. It is not uncommon that what turns out to be a suboptimal but acceptable design was used as the basis of the Feasibility Study with further optimisation undertaken following project approval.

A key component of a successful Feasibility Study is a well-considered and articulated study charter and Pre-Feasibility Study. Typically, a company will engage a third-party mining consultancy firm to prepare the Feasibility Study, and they may engage further specific subject matter experts. It is important to ensure the scope is well defined and that the third-party contractors' work is conducted in accordance with the study charter and prior studies. It is advisable that companies consult with specialists in their respective areas; for example, involve a metallurgist in reviewing the metallurgical test work and a mining engineer when assessing the capital expenditure requirements. It is also important to consider the ramifications of disclosures of the NPV of cash flows that are included in a Feasibility Study and before such disclosures are made, consult with a valuation practitioner to ensure the basis and parameters of the valuation is appropriate. This is detailed further in subsequent subsections.

It is encouraged that the preparers of the Feasibility Study and an independent third party critiques the Feasibility Study from the perspective of 'where would someone attack this?'. Conducting this review is important in preparing a robust study that can withstand the rigours of due diligence. As stated above, the purpose of a Feasibility Study can vary, but regardless of the purpose, it is likely that the assumptions of a Feasibility Study will be challenged. If the Feasibility Study is not supportable and is not able to stand up to due diligence enquiries, then the investment in the Feasibility Study has likely not been worthwhile.

SELECTING AND UNDERSTANDING PROJECT HURDLES AND METRICS

A clear and logical decision pathway for a project relies on selecting appropriate measures on which to assess the project. There is no single measure of a project that will give all the answers on whether a project should proceed and so it is important to consider a number of different hurdles and be aware of the benefits and limitations of each.

Net present value

The NPV of a project is one of the primary measures of whether a project is economically viable. It involves forecasting the expected cash flows of the project and discounting them back to present value using a discount rate that reflects the required rate of return. Broadly and simplistically, if the NPV of a project is greater than zero, then the project is considered economically viable.

The NPV of a project requires forecasting of cash flows and an assessment of the risks of those cash flows. The risks of forecast cash flows being achieved can either be reflected in the discount rate or by risk adjusting the cash flows. This is explored further throughout this chapter.

Internal rate of return

The internal rate of return (IRR) is an alternative way of expressing the NPV of a project as it still requires a forecast of the cash flows of the project. The difference between the NPV and IRR of a project is that the IRR represents a rate of return that equates the NPV of a project to nil. Therefore, if the IRR is greater than the project's required rate of return then it is an economically viable project. The IRR should not be used in isolation as it has a number of limitations including:

- it is a tool to assess whether a project is economically viable and should not be used to compare different projects over different durations
- it is not an appropriate tool to decide between two mutually exclusive projects, which is often the case when there are capital constraints and alternative mining projects being considered
- it ignores the funding requirement for a project and the possible dilution of existing shareholders' interests
- it ignores the dollar value of the project, ie a project may have a NPV of $100 million with an IRR of 20 per cent, compared to a project with an NPV of $10 million and an IRR of 30 per cent, and so if the decision is based on the IRR alone, a suboptimal decision could be made.

The IRR should therefore be used in conjunction with the NPV and can provide a useful headline metric for reporting to the Board of Directors and/or the market.

Payback period

The payback is a simple tool for estimating the period over which the upfront capital expenditure for a project is recovered. Similar to the IRR, it should not be relied upon in isolation but can provide useful information for management as it provides a simple assessment of the expected length of time the project will take to recover the upfront capital expenditure. The benefit of it being simple and easily understandable is also its limitation in that it ignores a few fundamental parts of project evaluation including:

- time value of money
- risks of the underlying cash flows
- project cash flows after the repayment period, therefore sustaining capital costs and rehabilitation is often not considered.

Cost metrics

A key driver of the value of a project, and one that is largely within the control of management, is the cost structure for the project. As we detail below, the value of a project is often most sensitive to external factors such as commodity prices and exchange rates; however, Boards of Directors and management teams are typically judged on the project or company-specific value drivers. Therefore, the capital and operating costs are important metrics to forecast and monitor over the life cycle of a project.

Different commodities will have different terminology when it comes to cost metrics, but using gold as an example, the World Gold Council standardised cost measures by introducing the all-in sustaining cost (AISC) and all-in costs (AIC) metrics in 2013 (World Gold Council, 2013).

AISC is the cost of sustaining current mining operations and comprises total cash costs, plus corporate costs, reclamation costs, sustaining exploration and study costs, sustaining capital exploration and development, and sustaining capital expenditure. AIC refer to the aforementioned costs but includes the above costs incurred in expanding production (rather than just sustaining current operations).

These measures were introduced to provide all stakeholders with transparent and comparable metrics that reflect as close as possible the relevant cost of producing and selling an ounce of gold. The concept of AISC and AIC originated with gold but has been extended to other commodities. There are no hard and fast rules as to which costs should be included or excluded from these benchmarks; however, consistent with the key theme of the chapter, transparency is key. The users of the study will evaluate the company's project costs against forecast commodity prices as well as against peer companies that are already producing. This is particularly relevant for commodities that are less scarce in nature, with the cost per unit of production being one of the key factors considered when conducting an economic evaluation of the project.

APPLICATION OF EXTERNAL FACTORS

The level of rigour placed into the assessment of external factors affecting the economic viability of a project is typically not commensurate with the order of magnitude that these impacts can have on the value of a project. Admittedly, by their nature, these external factors are beyond the control of management, with every market participant affected by varying degrees to changes in these factors. However, as a sensitivity analysis often reveals, the value of a project is typically most sensitive to external factors such as commodity prices or exchange rates. We set out below an overview of the external factors that should be considered when evaluating a project and how these should be considered when assessing the economics of a project.

Commodity prices

There is a widely varying level of attention given to the forecast commodity prices used in the preparation of mining study models. A key theme to the treatment of commodity prices though is that there is not enough emphasis placed on the reasonableness of forecast pricing assumptions and the basis for the pricing assumptions made. Some mining study models may be as rudimentary as using the current spot price and using that indefinitely over the life-of-mine.

One common approach is to use subscription services that present consensus pricing in the industry. However, companies should exercise caution in relying solely on consensus analyst views as the views of analysts may contain significant bias. Analyst forecasts can be readily sourced for those commodities that have benchmark pricing and significant analyst coverage. It is important for companies to consider the rights attached to subscription services and discuss publishing rights with data providers ahead of releasing study results publicly.

However, using market research and a consensus view on pricing can be more challenging when:

- the commodity characteristics are not homogenous and there is not a benchmark price that can be used to estimate a realised price
- there is not a deep spot market for trading in a particular commodity
- the commodity is traded through long-term contracts, the terms of which may not be publicly available.

In situations like the above, recommended approaches to assessing forecast pricing could be:

- Where realised pricing is expected to differ from the benchmark price, and forecast analyst benchmark pricing is available, consider the use of historical pricing relative to the benchmark pricing and use this to infer forecast-realised pricing over the life-of-mine.
- Where contract pricing is used, test the market. Approach potential purchasers and negotiate non-binding term sheets for an offtake. This option is typically only available at the Feasibility Study phase as it is unlikely that a potential offtake partner will negotiate at the Pre-Feasibility Study or Scoping Study phase of a project.
- Consider the economics underpinning that particular commodity. For example, what are the global demand drivers for the commodity and what does the supply landscape look like? This can help inform the company's view of forecast pricing when it comes to project evaluation.

Throughout this chapter, we discuss the messaging of the business case and the importance of ensuring that the study charter aligns with the strategy, goals and objectives of the company. It is equally important that the economic inputs to the model, such as commodity pricing, is consistent with the story of the study. For example, the study should include an analysis of the industry, the drivers of which should align with the business case and be reflected in the company's forecast pricing.

In preparing the mining study model, management should align the commodity price assumptions with the risk tolerance set at the Board of Directors level. For example, if the company's risk register provides for mitigation of commodity price risk through use of hedging instruments then the cost of hedging and the expected price received should be modelled.

Exchange rates

Similar to the discussion above on forecast commodity prices, the value of a project can be highly sensitive to forecast exchange rates, which can significantly affect the viability of the project. This is common for projects based in Australia as all costs are incurred in Australian dollars, yet most commodity pricing is denominated (and commodities transact) in US dollars. Therefore, if the Australian dollar appreciates relative to the US dollar, the Australian dollar commodity price falls, which reduces revenue and profitability of a project. Companies are not expected to be able to predict the exchange rate with any degree of certainty, but it is expected that companies are aware of their foreign currency exposure and take a position according to where they might reasonably expect it might go.

As has been touched on many times throughout this chapter, it is important that the risk tolerance of companies and their views of foreign currency movements are reflected in the model. This should either be in the forecast foreign exchange rate that is used or if it is a material risk that is identified in the studies, then the use of hedging could be used in the model. Mining projects may have a natural foreign currency hedge, where the costs and revenue are denominated in the same currency (as is often the case for projects in the United

States or projects in counties where capital costs are denominated in US dollars). Other hedging options that could be considered by companies that are at the Feasibility Study / development phase are foreign currency options or the use of forward or futures contracts.

Taxation

Tax tends to be either oversimplified or even ignored when preparing mining studies and mining models. This subsection does not purport to be a complete tax guide as the individual tax circumstances of a company will differ on a case-by-case basis and will depend on the jurisdiction in which it operates. However, in order to assist companies in determining whether or not a tax issue may exist, set out below are a few areas of tax that companies should be aware of:

- What is the tax jurisdiction of the project and how is it taxed? Is there a withholding tax associated with repatriating profits to Australia?
- Consider the prevailing tax rates of the company's tax jurisdiction and expected changes in the applicable tax rates over the forecast period.
- It is common for mining companies to have been loss making while in the exploration and development phase, therefore typically mining companies will have carry forward tax losses that may be applied to future profits. This means that a part of a company's forecast profits may be tax free. Utilisation of these losses are conditional on the passing of either the similar business test or the continuity of ownership test. As the rules around the use of tax losses are complex, it would not be appropriate to assume the availability of such losses without some input from a tax specialist.
- Tax depreciation is different to accounting depreciation and this will give rise to deferred tax assets or deferred tax liabilities. This distinction forms an important part of the three-way financial modelling process.

It is advised that companies seek specialist tax advice when evaluating a project. If the Feasibility Study is being provided to potential investors as part of due diligence, this can be an area that is challenged and so unsupportable tax assumptions may affect the credibility of the model.

Environmental factors and permitting

Guidance on economic evaluation of mining projects would not be complete without mentioning the impact of environmental factors and permitting. An aspect of the economic evaluation of mining projects that is becoming increasingly important is the preparation of an environmental impact assessment (EIA). The EIA process helps identify the possible environmental effects of a proposed activity and how those impacts can be mitigated (Environmental Law Alliance Worldwide, 2010). It is an important aspect of economic evaluation because if the environmental impacts of the project are not able to be mitigated to an acceptable level then project finance will become increasingly difficult to obtain, which is a significant barrier to project development and production.

In the context of valuing a mining project, the environmental risks and risks around permitting and tenure can be reflected by risk adjusting the forecast cash flows of the project or through an adjustment of the discount rate. This is relevant to the life cycle of a mining project because an equity investor will value the project as part of its decision on whether to invest. If the environmental risks and risks around tenure are not mitigated to a level such that the risk adjusted cash flows supports an investment, or the discount rate is so high that the project is not economic, then the equity investment will not proceed. The impact of environmental, social and permitting factors are detailed in Chapter 2.

SELLING THE COMMODITY AND OTHER MARKET FORCES

Consideration of the external factors set out above is a crucial part of economic evaluation. However, it is also important to consider the logistics and economics of selling the commodity. We have set out below a non-exhaustive list of areas that should be considered initially in the study charter and considered with additional rigour as the project progresses through the study phases.

Customer and terms of supply

As detailed in the assessment of forecast commodity prices, a study should consider the economics of the commodity, including demand and supply drivers. In conducting an economic evaluation of a project it is also important to consider the end user of the commodity. Questions that may require answering include:

- Will the commodity be exported or used domestically?
- If it is to be exported, has shipping and freight been considered? Is the expected pricing on a free on board; cost and freight; or cost, insurance and freight basis?
- Will the supply of the commodity be contracted over a fixed term, eg an offtake. If so, what are the expected terms of this offtake? At the Scoping and Pre-Feasibility Study phases, a number of scenarios can be presented; however, the company should have a clear sales strategy mapped out in the Feasibility Study.
- Will the supply of the commodity be on an exclusive basis or can the company also source alternative buyers?
- Will there be treatment and refining charges payable to the smelter? If so, this should be reflected in the project model to a sufficient level of accuracy depending on the study that is being conducted.
- Are there penalties that are payable to the buyer for quality issues?
- Are there royalties payable on production? If so, this should be reflected in the model. Historical and government royalties over a company's project should be known at the time of conducting a Scoping Study and therefore should already be included. However, if the company intends to issue a royalty as part of a financing package then this should be included at the Feasibility Study phase.

SENSITIVITY ANALYSIS AND SCENARIO ANALYSIS

Sensitivity analysis

A sensitivity analysis is an important tool for demonstrating the range of possible valuation outcomes for a mining project. The quantum of the sensitivities applied should reduce as the project progresses through the various study phases. This is because as the level of confidence of the estimates improves, the range of possible outcomes should narrow. For example, the sensitivities applied to a Scoping Study will be higher than those applied in a Feasibility Study. Not only does a sensitivity analysis show the user of the study a range of possible outcomes, it provides valuable insight into which inputs have the greatest effect on value. This can aid in the risk assessment of a project and how a company seeks to mitigate the key risks.

The inputs to a mining model that we would recommend typically be sensitised are as follows:

- forecast pricing
- assumed exchange rates (where applicable)
- unit operating costs, for example mining and processing costs
- upfront capital expenditure
- sustaining capital expenditure
- rehabilitation costs
- grade and recovery assumptions
- timing and level of production.

Scenario analysis

A scenario analysis is the identification of possible events or scenarios and evaluating the outcome of those scenarios occurring. This is typically in the form of assessing the impact of these events or scenarios on the NPV of a project. This differs from a sensitivity analysis as it involves assessing the impact on an output (usually NPV) for particular events occurring, compared to a sensitivity analysis, which shows the impact on an output (also typically NPV) for changes in an input. A scenario analysis can involve a change to a number of different inputs at the same time, whereas a sensitivity analysis is often applied to inputs in isolation (noting that sensitivity matrices can be used).

A scenario analysis is often used at the Pre-Feasibility Study phase in order to decide on which of the proposed alternatives is the optimal one to proceed with. Although it is suited to a Pre-Feasibility Study, it can also be used in Scoping Studies, albeit to a lower level of confidence.

The benefits of conducting a sensitivity analysis are that it can enable a better understanding of the uncertainties of a mining project and can allow management to develop strategies to deal with certain scenarios or events, should they occur.

Examples of scenarios considered in the evaluation a mining project may be:

- invest additional funds upfront to allow increased throughput capacity at the company's processing plant
- the company uses a contract miner to operate its mine as opposed to sourcing its own staff and equipment
- the company only mines the ore that is accessible by open pit mining
- the company stores and processes tailings
- if environmental approvals are not received and therefore timing of construction is delayed
- scenarios around haulage and transport, eg construct a haul road or invest in rail.

ASSESSMENT OF THE MOST APPROPRIATE AND CONSISTENT PRESENTATION BASIS

A common pitfall of financial models is that the preparers of the model have not considered the basis on which the forecast is prepared. Set out below are a few factors that should be considered when preparing a financial model for a mining project. The Australian Securities and Investments Commission (ASIC) and ASX provide guidance on the nature and extent of what a company can disclose to the market about the results from mining studies depending on the phase of the study. Of particular relevance is ASIC's *Information Sheet 214: Mining and Resources – Forward-looking Statements*. Those looking to disclose the results of a study should make themselves familiar with the guidance contained in this document before doing so.

Further, the release of a NPV of a project can put a stake in the ground in terms of the market's expectation of value. In the event that the company's project is the subject of a transaction that requires an independent expert's report, for example, for a related party transaction, takeover or other control transaction, then an independent valuation specialist will opine on the value of the project, which due to an inconsistent basis of valuation, may be materially different to the valuation results announced as part of the study. Further, when/if the company reaches production there may be significant variance in values between the assessed value in the study, and the values in the project model, which for financial reporting purposes, the company's auditor would review for impairment. Again, this may have commercial implications as the market is unlikely to view an impairment in the early phase of production favourably.

Nominal versus real cash flows

In preparing mining models, it is often overlooked as to whether the cash flows are forecast in real terms or nominal terms.

Real cash flows are cash flows that are forecast in today's dollars, which in effect assumes that an inflation rate of nil is applied to future revenue and costs. Whereas, nominal cash flows are cash flows that are forecast into the future in future dollar value terms, based on expected inflation over the forecast period.

Historically, in a low inflation environment, the impact of inflation may not have been material to the value of a company's project. However, with inflation rising globally, it has become increasingly important to understand the impact of inflation when forecasting costs and revenue.

It is worth noting that the value of a project should be equivalent whether the model is prepared on a real basis or nominal basis; however, in practice, the issue arises when the model does not treat nominal and real cash flows consistently. Further, it is common for there to be a mismatch between the cash flows and the discount rate; for example using real cash flows but a nominal discount rate, or for not all revenue and costs to be treated on a consistent basis. The impact of this can be:

- it can distort the value that is announced by the company publicly
- it can impact internal decision-making if the value and required rate of return has not been assessed appropriately
- if all revenue and costs are not forecast on a consistent basis, it can affect economic decision-making by the company
- if all revenue and costs are not forecast on a consistent basis, it can impact the reliability of forecasts and the level of reliance placed on them by external parties conducting due diligence on the project.

The notion of real and nominal cash flows is more commonly understood to apply to the costs of a project; however, it should also be considered when forecasting commodity prices.

Discount rate

The discount rate used for a financial model in a mining study should represent the required rate of return of the business or company. Therefore, the most appropriate discounted rate to use for the assessment of the value of a project would be the weighted average cost of capital

(WACC). A component of the WACC is the cost of equity, which is typically derived using the capital asset pricing model (CAPM). The CAPM is commonly used in determining the market rates of return for equity investments and project evaluations. In determining a business' WACC, the CAPM results are combined with the cost of debt funding. It is worth noting that the cost of debt, cost of equity and the funding structure of a project is typically not known at the time of completing the studies; therefore, these assessments require a best estimate at the time and can be refined as the project progresses through the study phases. WACC represents the return required on the business, while CAPM provides the required return on an equity investment. In assessing an appropriate discount rate for the project, the authors of the studies should be aware of the difference between the discount rate for a company or business compared with the discount rate of an asset or project. In practice, a company discount rate is often used as a proxy for the asset-specific discount rate.

In light of the above discussion on nominal versus real discount rates, it is important to note that the commonly used formula for the WACC (set out below) represents a nominal post-tax discount rate:

$$\text{WACC} = \frac{E}{E+D} * Ke + \frac{D}{E+D} * Kd * (1-t)$$

where:

WACC = weighted average cost of capital

E = value of equity

D = value of debt

K_e = cost of equity

K_d = cost of debt

t = tax rate

It is common for mining models to assume straight-line pricing over the life-of-mine model, which typically represents a real forecast. Therefore, if the above formula is used for the assessment of the discount rate and real pricing is adopted in the model, then there is misalignment between the discount rate and forecast cash flows. This equally applies to forecast costs.

In practice, there is often not much science that goes into a discount rate assessment for a mining study and a rule of thumb rate of say ten per cent is applied. The risk of using such a rule of thumb is that when a more thoroughly assessed discount rate is applied at a later date, decisions made based on values derived when using the rule of thumb may be found to have been wrong. Errors in this area can distract away from the key messaging of a study or reduce the level of reliance placed on it by users of the study. Further, it is common for companies to use a rule of thumb rate for internal decision-making when assessing competing projects; however, this typically has minimal relationship to risk and may lead to suboptimal decision-making.

When preparing a study, caution should be exercised in selecting a discount rate. A peer company may use a particular discount rate but often the disclosure is not sufficient to allow the reader to determine the extent to which the discount rate reflects the specific risks of the project or whether the risks of the project are reflected in the cash flows. Therefore, just because a particular discount rate is used by a peer company, it may not be appropriate to use for your project.

Funded versus unfunded

The value of a project will differ depending on whether the valuation is predicated on the assumption that the project is funded or whether it is subject to funding. Essentially this reflects the difference between the value of the project itself and the value of the project to the company and its shareholders. As detailed in ASIC's *Information Sheet 214: Mining and Resources – Forward-looking Statements*, there is no specific requirement for funding to have been obtained in order to publish production targets or forecast financial information. However, there is the legal requirement that a company be able to demonstrate it has 'reasonable grounds' that it could obtain the requisite project finance as and when required.

INCORPORATING THE REQUIREMENTS OF PROJECT FINANCIERS

Project financing can come in many forms; however, a detailed discussion of these financing options is outside the ambit of this chapter on economic evaluation of mining projects. However, set out below are the more common project financing options that are used by industry participants:

- equity placement to a cornerstone investor
- initial public offering of its shares on the ASX or other securities exchange
- project finance and other forms of loan funding from banks or other financial institutions
- convertible debt, typically in the form of a loan that is convertible at the election of the financier such that they are exposed to the potential upside of the equity, but can rank ahead of equity holders in the event that the project does not generate sufficient economic returns
- corporate bonds
- joint venture and earn-in agreements
- mezzanine finance, which can be provided by parties seeking to secure offtake
- streaming facilities, whereby companies repay the amount drawn by delivery of the commodity at a discount to the market price
- royalties, typically on a net smelter return basis.

Each of the above forms of equity, debt and hybrid debt financing involves sharing of risk between the financier and the company. Since the global financial crisis, where loan terms were up to 20 years, we have seen a transition away from banks providing longer term finance to cover the period of development and production, to generally shorter-term loans of five to seven years.

In the subsection titled, 'Expressing the output from the preceding chapters in the form of a financial model', we introduced a few key elements that a potential investor or debt financier will expect to be available when undertaking its due diligence. In working toward the ultimate goal of a profitable producing mine, it is important that when building models and completing the mining studies, companies plan for the project financing phase of a mining project.

The requirements of a financier will differ on a case-by-case basis, depending on a number of factors including but not limited to the risk appetite of the financier, the type and quantum of funding proposed, the technical, environmental and economic risks of the project and the level and quality of the security granted. Set out below are examples of covenants or metrics that financiers may require initially and then on an ongoing periodic basis:

- gross leverage ratio, which is typically calculated as total debt / earnings before interest, taxes, depreciation and amortisation (EBITDA)
- interest cover ratio, which is typically calculated as EBITDA / interest expense paid over the last year
- net gearing ratio, which is typically calculated as total net debt / (total net debt + equity)
- reserve tail ratio, which is typically calculated as (remaining Ore Reserves to be mined at a specific date / total Ore Reserve)
- maintaining a minimum level of earnings or free cash flow.

In addition to the above covenants, a project financier will typically require an independent third-party review opinion on the financial model. There are a number of key pillars to an effective financial model, which are summarised in 'Expressing the output from the preceding chapters in the form of a financial model'. In addition to the discussions of this subsection, it is best practice to consider the likely scope of an independent third party that may ultimately review the model. The scope of the review opinion will differ on a case-by-case basis depending on the project, the study phase and the risk tolerance of the financier. However, set out below is a summary of the scope that a financier may require from an independent third party when reviewing a financial model:

- the internal logic and mathematical formulae incorporated in the model and that the various calculations are sound, correct and internally consistent and that the accuracy of the results can be relied upon
- the model logic and calculations are consistent with and correctly reflect the terms, parameters and mechanisms set out in agreements covering offtakes, finance including cash sweeps and covenants, contract mining terms, streaming facilities, hedging and/or other commercial agreements
- formulae are consistent across all time periods and there are no circularities
- the model allows changes to assumptions to flow correctly and logically through to the results
- the model has been suitably constructed so that, after adjusting the base case assumptions and input data for sensitivity analysis, the results are accurate, logical and can be relied upon
- the input data used in the model is consistent with supporting documentation
- input data and calculations in the model are consistent with the extracts from supporting project documentation
- any macros that are used to govern the calculation of the model are correct and can be relied upon.

CONCLUSIONS

This chapter provides guidance on best practice when conducting an economic evaluation of a project. While the level of rigour of the economic assessment will increase as the company progresses from a Scoping to a Pre-Feasibility and then a Feasibility Study, it is important to have strong foundations with an eye to additional detail at later phases. This will mean that should the project proceed to Pre-Feasibility and Feasibility, a company will not be required to 'reinvent the wheel' as the company would have the requisite financial model and framework in place, albeit to a lower level of confidence. This also demonstrates the importance of having a study charter that adequately addresses the company's position on risk and the alignment of the study to the company's strategy. It is not uncommon for mining studies to be prepared for a particular purpose, ie to obtain funding, therefore sometimes these can be set and forget once that purpose is achieved. However, the most effective study charters and subsequent mining studies are live documents that are referred to as the company progresses through to production. This can ensure that the company's operations remain in line with the company's strategy and objectives, and where deviations to the strategy or risk tolerance are observed, the reasons for these deviations or amendments can be explained to key stakeholders.

REFERENCES

Environmental Law Alliance Worldwide, 2010. *Guidebook for Evaluating Mining Project EIAs,* first edition, pp 4, 19 (Environmental Law Alliance Worldwide: Eugene, OR).

JORC, 2012. Australasian Code for Reporting of Exploration Results, Mineral Resources and Ore Reserves (The JORC Code) [online]. Available from: <http://www.jorc.org> (The Joint Ore Reserves Committee of The Australasian Institute of Mining and Metallurgy, Australian Institute of Geoscientists and Minerals Council of Australia).

World Gold Council, 2013. All-in sustaining costs and all-in costs [online]. Available from: <https://www.gold.org/about-gold/gold-supply/responsible-gold/all-in-costs>.

RISK

11

INTRODUCTION

A project progresses though the following phases of development prior to its execution:

- Scoping Study
- Pre-Feasibility Study
- Feasibility Study.

These phases have specifically been designed to provide decision-makers and investors with the required information and confidence of expected outcomes at key decision points or tollgates prior to commitment and allocation of capital to execute the project. Stakeholders require the following risk information when project approval decisions are made:

- What are the project risks associated with each phase of the project?
- For each identified risk, what is the probability that the risk could occur and what are the consequences if the risk did occur?
- What are the material risks that could impact on business objectives?
- What are the key controls and cost to mitigate or manage material risks?
- Is the overall project risk profile acceptable and are the material risks within predetermined stakeholder tolerability limits?

Risk management therefore plays an important role to ensure the project business case presented is understood and communicated appropriately to all stakeholders. Risk management should be undertaken in all phases of a project, from the initial Scoping Study through to execution. The risk profile of the project will change as the project matures. Thus, risk management shall be a continual activity during every phase of project development.

The risk chapter aim is to provide the key risk management processes that need to be focused on during each of the project development phases, and thus provide stakeholders with the required level of confidence to make informed and effective decisions. These decisions may include:

- approving the project as presented
- determining if more work is required to provide an acceptable project risk profile for the stakeholder investment
- determining that the project risks are too great and cease continuing work on the project.

RISK MANAGEMENT PROCESS

Risk is the effect of uncertainty on objectives according to the definition provided in the international risk standard ISO 31000:2018.

Identification of the uncertainties affecting project scope, definition, cost, schedule, contracting strategy and execution workplan determines the success of a project. Risk management is therefore the art and science of identifying, analysing and responding to risk uncertainties throughout the life of a project to ensure it meets its objectives. Failure to perform effective risk management can cause a project to exceed costs and/or schedule or not achieve the expected performance targets, hence not achieving its business case and objectives.

Risk management is a necessary consideration each time a decision is made. Projects must align activities and decision-making with outcomes that help reach the business objectives.

The risk management process is the series of defined steps by which risk is identified, assessed and managed. For this process to be effective, it is important to ensure that there is participation during risk workshops from a diverse group of people, whose combined knowledge will cover all of the risks that are likely to have a material impact on the project outcome.

Risk identification

Risk identification is the structured process of identifying all the risks, issues and opportunities relevant to the project and documenting these in a comprehensive risk register. A risk description should describe the risk event succinctly and how it could impact the project objectives. This should include where, when, why and how any such event or effect could prevent, degrade, delay, accelerate or enhance the achievement of project objectives.

The effectiveness of the risk management process is reliant upon the quality and breadth of the collection of risks identified during each phase of development. The comprehensive identification of risks within a register is critical, as any risks that are not identified will not be managed. Any risks that may affect the business objectives but are not under the control of the project or organisation must also be included so that mitigation strategies or contingencies can be developed and the external impact is understood.

Analyse risk

Risk analysis is the process of reviewing each of the identified risks to determine the causes of that risk, its likelihood of occurrence and the consequence(s) if it did occur.

Causes for each risk are identified to understand the potential causal pathways for that risk, and hence ensure that the selected risk controls will effectively manage these causal pathways. The risk analysis involves consideration of the positive and negative impacts and the likelihood that those impacts may occur by determining risk sources, risk scenarios and controls.

When identified risks are analysed for their likelihood of occurring and their consequence of occurring, this assessment must be individually made against those different risk impact types that are likely to affect that particular risk. These risk impact types are typically health/safety, environment, community, legal (regulatory), reputation, financial or similar. The risk impact types ensure business objectives are considered on a holistic basis. This process generally uses qualitative information as part of risk assessment workshops.

Note: Where information is available relating to cost and schedule impacts for each risk, it is beneficial to perform quantitative analysis to inform a decision on the appropriate cost and schedule contingency. This is normally carried out by finance personnel as part of the financial analysis using tools such as a Monte Carlo simulation.

The likelihood of a risk occurring, and the extent of its consequences, are combined to estimate the level of risk associated with that event using defined assessment criteria, which will vary by organisation depending on their risk framework, materiality levels and risk appetite.

When a risk has been identified as being a material risk from the defined assessment criteria, it is recommended that the risk bow-tie methodology is applied to that risk, to help identify the critical controls required to effectively manage it. A risk bow-tie methodology is a visual risk management technique used to assess and analyse the potential risks associated with a particular event, process, or activity. The primary purpose of a risk bow-tie analysis is to identify and understand the various factors that can lead to the occurrence of a risk event, as well as the potential consequences that may result from it. This methodology assists the project to gain insights into the critical risk pathways, control measures, and mitigation strategies needed to prevent or minimise the impact of the risk event.

Evaluate risk

Risk evaluation is the process used to define the seriousness of the risk in relation to other risks. It uses a predefined company risk matrix to rank the project risks based on priority. Understanding the project risks priorities then allows:

- resources to be allocated in the most effective manner, to the most important (material) risks
- those risks that don't meet predefined stakeholder tolerability limits to be identified, which then ensures that either additional controls are in place to reduce the risk level, or where this is not practical, for stakeholders to make informed decisions relating to the project's future.

Treat risk

Risk treatment is the process of identifying, selecting, developing and implementing risk controls to modify a risk to an acceptable level. It also includes assessing the effectiveness of the risk controls implemented. Effective risk controls need to be:

- able to reduce either or both the likelihood of a risk occurring or the impact of the risk
- practical to implement
- located as high as possible in the hierarchy of controls (ie reliable)
- timely (it operates at the right time and quickly enough)
- cost-effective
- able to effectively manage one or more of the identified risk causes.

For risks that have been identified as being a material risk (see 'Analyse risk' section), it is important that the critical control is identified. A critical control is a control that is crucial to preventing the event or mitigating the consequences of the event. Or put another way, the absence or failure of a critical control would significantly increase the risk despite the existence of the other controls. A simple example would be faulty brakes in a car.

Monitoring and review

The monitor and review process must provide stakeholders sufficient insight on the performance of risks and controls to consider whether further action is required. The management of risks is continual and involves regularly and systematically reviewing and reassessing risks, checking the effectiveness of risk treatment plans and recording and reporting progress.

Risks should be monitored from both an external and internal context to ensure changing circumstances do not alter the risk assessment or priorities, or introduce new risks to the project. Risks will not remain static and therefore, the risk management process needs to be regularly repeated, so that new risks are captured in the process and effectively managed.

The single source of risk data information must be contained and continually updated in the risk register, with a single person responsible for maintaining the integrity of the risk register.

Communication and consultation

Sound communication with stakeholders will establish expectations, manage perceptions and ensure that their needs are considered while they are kept informed of project developments. Effective communication and consultation are essential to ensure that those responsible for implementing risk management, and those with a vested interest, understand the basis on which decisions are made and the reasons why particular treatment options are selected. This is achieved by:

- bringing different areas of expertise together for each step of the risk management process
- ensuring different views are considered when defining risk criteria and evaluating risks
- providing sufficient information to facilitate risk oversight and decision-making
- building a sense of inclusiveness and ownership among those affected by risk.

It is particularly important that both senior management and members of the Board of Directors understand the key (material) project risks that could have a serious impact on the project objectives should any of them eventuate, and the effectiveness of the controls managing these risks.

KEY RISK MANAGEMENT DELIVERABLES FOR PROJECTS

The key risk management deliverables for any project as a minimum should be:

- a formal risk register
- end of phase risk chapter
- control actions included in forward workplan
- the risk management plan.

The detail and key focus area for each of the above deliverables will vary during each phase of the project life cycle and will be discussed in more detail in the project life cycle section of this chapter.

Risk register

The risk register can be maintained in any format (ie Microsoft Excel, Microsoft Word, data basis, application software, etc), but is essentially a document that records the project risks and includes appropriate information about each risk to allow the assessment of the risk and identification of controls to manage the risk to an acceptable level. The essential fields that should be included in a risk register are:

- *Risk description* – The risk description should be generated detailing the relationship between the causes, event and impacts. The objective of this statement is to provide stakeholders with a quick snapshot of the risk in a clear and concise statement. A method of capturing a risk in a simple sentence is to use a structure such as the following:

[**Risk event**] caused by [**cause or trigger of the risk event**] resulting in [**impacts**].

- *Causes* – The causes are the conditions leading to a risk event or that give rise to uncertainty.
- *Impacts/consequences* – The outcome or impact of a risk event should it materialise. Impact or consequence should not just be based on cost, schedule or performance, but consider impact categories such as safety, financial, environment, legal and reputation. The impact or consequence should be tied to a key project deliverable or activity to better articulate the concern.
- *Controls* – Controls must address the cause of the risk event and must be a measure or action that modifies risk by either preventing the risk from occurring or mitigating (reducing) the consequence of the risk. Critical controls that are identified should be visible in the risk register.
- *Current risk rating* – The risk rating based on the effectiveness of controls currently in place. This is important to understand the current risk exposure and to determine if further actions or mitigations are required.
- *Actions/mitigation* – These are proposed controls to further reduce the risk exposure. Actions should reduce risk to as low as reasonably practicable (ALARP) and there should be a direct relation between actions in the risk register and the forward workplan. Actions should further follow the SMART principles, being:
  - S – Specific
  - M – Measurable
  - A – Agreed
  - R – Realistic
  - T – Time-bound.
- *Target risk rating* – The risk rating based on the effectiveness of controls currently in place and those proposed to be implemented. This should provide the end target risk profile that stakeholders should consider in determining if the risk can be managed to an acceptable level and within predefined risk tolerability limits for the business.

The risk register should further reflect the risks included for key activities and project deliverables to demonstrate risk integration across project disciplines. Some important aspects to consider to assist in identifying risk that could affect project objectives are discussed below.

Consideration of project deliverables could assist in creating an initial list of risk inventory.

End of phase risk chapter

The end of phase risk chapter can be either a standalone chapter or can be incorporated into the risk section of the overall study report. The purpose of the end of phase risk report is to provide an overview of the risk activities during the specific phase, communicate the current and target risk profile of the project to stakeholders and clearly state the actions and cost required to achieve the target risk profile.

The end of phase risk chapter must align with the investment approval request and what is communicated to the stakeholders, decision-makers and/or investment committee.

The typical topics to cover in an end of risk phase chapter are:

- project alignment to strategic risk
- risk identification activities and information
- risk register summary.

Project alignment to strategic risk

The project needs to demonstrate how the objectives align to the overall organisational strategy and business objectives. Examples of project objectives versus organisational business objectives are provided in Table 11.1.

There should be a clear alignment between the risk identified in the risk register and the project and business objectives. The risk impacting the business objectives should be specifically articulated to stakeholders.

Risk identification activities and information

This section should reflect the risk activities conducted during each study phase. A list of workshops conducted during the phase and reference to other sources of information considered such as hazard operation (HAZOP) reports, failure mode and effects analysis (FMEA) studies, safety in design (SiD) documents, etc.

Risk register summary

Material risk should be discussed in detail to understand the drivers and impact of the risk as well as the proposed risk treatment actions, the timeline and cost associated with the risks.

A general summary should then be provided for the overall risk profile reflecting the number of risks per category to provide stakeholders with a general overview of the granularity of the risks identified.

Forward workplan

The forward workplan should be the mechanism to ensure that risk mitigation actions identified in the risk register to reduce unacceptable current risk ratings to desired target risk ratings, are included and budgeted for in the next phase of the project. There should

TABLE 11.1

Risk and business objectives.

PROJECT OBJECTIVE	ORGANISATIONAL BUSINESS OBJECTIVE/STRATEGY	EXAMPLES OF PROJECT RISKS
Develop infrastructure capable of providing sustainable supply chain capacity of XX Mtpa.	Grow and increase supply chain capacity over the next five years to XX Mtpa.	Plant name plate capacity not achievable caused by lack of sustainable maintenance strategy. Process bottleneck reduces plant throughput.
Design and engineer infrastructure to required international standards.	Maintain asset integrity and performance of fixed plant infrastructure to eliminate production interruptions.	Failure of girth gear due to design fault.
Consider safety in design for infrastructure development.	Zero harm to employees and contractors.	Loss of containment.
CO_2 emissions are minimised though design and project procurement.	Target zero emissions by 2050.	Increased exposure to CO_2 due to increased production capacity.

be a clear one-to-one relationship between the risk register and the forward workplan. It is therefore expected that an action connected to a material risk in the risk register will be specially included in the forward workplan. If, for example, a risk has been identified that the plant throughput could not be achieved due to process bottlenecking issues and an action was identified to conduct a debottlenecking study, there should be an action in the forward workplan to conduct a de-bottlenecking study that has an owner, date and cost assigned to it.

The number of risks and control actions assigned to a single risk owner/control owner needs to be realistic, particularly when managing material risks, to avoid overloading that individual. This is important as a poorly managed material risk and the associated controls (in particular the critical controls) could likely result in the project being impacted by that risk!

Risk management plan

The risk management plan is often confused with a risk mitigation plan. The risk mitigation plan is essentially the risk action plan included in the forward workplan, where a risk management plan deals with the risk methodology applied and the risk management activities of the project.

This risk management plan is the principle reference for the risk management activities conducted of the project. It defines how the project team coordinates risk activities by establishing and communicating standards down to the project level then collating and reporting on risks in a bottom-up approach. It includes resources assigned to assist with the management of risk and to encourage a consistent and comprehensive language and approach to risk management across the project. It also provides details of the principles, framework and process to manage risks within the project.

The key elements that should be included in a risk management plan are:

- defining risk roles and responsibilities
  - risk lead / manager
  - risk owner
  - control owner
  - action / treatment owner
  - project lead / director
- schedule of risk management activities
  - establishing risk register
  - performing risk governance activities
  - undertaking cost and schedule quantitative analysis (finance team)
  - risk reporting
  - risk training
- risk management process
  - risk framework
  - risk assessment criteria
  - materiality and risk tolerability criteria
  - risk register repository.

The following section of this chapter deals with the specific risk activities, focus and deliverables for each phase of the development of a project.

PROJECT LIFE CYCLE – RISK DELIVERABLES

A summary of risk deliverables for each phase of the project study are provided in Table 11.2 and are further discussed in the remainder of this section.

Scoping Study

Objective

To understand the likelihood of a project achieving the desired business objectives prior to committing larger sums of money, by using a non-optimised Scoping Study level project option, to identify and understand the most significant risks (or fatal flaws) should it proceed to implementation.

Key deliverables

- Identify and analyse the material business risks associated with a non-optimised version of the potential investment opportunity.
- Develop an initial risk register for the non-optimised version of the potential investment opportunity.
- Document any material risk areas of intolerable high target or residual risk rating or potential fatal flaw. Identify and describe any control or treatment strategies to manage the risk to a tolerable level.
- Prepare the risk section for the Scoping Study report.
- Prepare a risk management plan for the Pre-Feasibility Study.

Note: If no cost-effective mitigation strategies can be deployed to remove fatal flaws or to reduce risk to an acceptable tolerance level, the investment decision should be reconsidered at this point before commencing any further study.

TABLE 11.2
Risk study phase guide.

RISK					
Activity	**Deliverable**	**Description**	**SS**	**PFS**	**FS**
Risk assessment	Risk register	Identify fatal flaws	✓		
		Identify material risks	✓	✓	✓
		Identify controls to reduce material risk exposure		✓	✓
Risk profile	End of phase risk chapter	Material risks impacting on business	✓	✓	✓
		Risk treatment/controls to reduce risk to an acceptable level		✓	✓
		Assurance on control effectiveness to reduce risk to acceptable level			✓
Risk management plan	Risk management plan	Define risk management methodology and activities for next phase of project	✓	✓	✓
Forward workplan	Forward workplan	Mitigation actions identified in the risk register to reduce material risk to an acceptable level are included and budgeted for in the next phase of the project	✓	✓	

SS = Scoping Study; PFS = Pre-Feasibility Study; FS = Feasibility Study.

Risk management tools and techniques

- *SWOT analysis* – Used to understand the strengths, weaknesses, opportunities and threats of the non-optimised Scoping Study level project option.
- *Bow-tie analysis* – Used to assess risks that have causes and consequences to identify controls to prevent causes or to mitigate or minimise the consequence. This may be useful in this early project phase to better understand any fatal flaw risk(s) that have been identified.

Risk lead accountabilities

- Agree scope of risk assessment based on project objectives with project lead.
- Agree on subject matter experts and functional resources to participate in risk assessment.
- Coordinate and schedule risk workshops.
- Conduct risk workshop(s).
- Capture risk outcome in agreed risk register template.
- Agree to risk ownership.
- Implement process for continual review and update of risk register until end of phase.
- Obtain risk register sign-off from project lead team.
- Draft the risk chapter for the Scoping Study report.
- Draft risk management plan for next phase.
- Agree on the risk management strategies to go into forward workplan.

Pre-Feasibility Study

Objective

To assess risks associated with each of the Pre-Feasibility Study options being considered and use this output as a key input in the selection process for preferred investment alternative.

Key deliverables

- To advise the project on most appropriate risk assessment technique to use to assist with the selection of preferred investment alternative.
- Identify and analyse the material business risk associated with the project scope and objectives.
- Develop the risk register in sufficient detail to support the selection of preferred investment alternative.
- Prepare risk section for the Pre-Feasibility Study report.
- Prepare a risk management plan for the Feasibility Study.
- Ensure risk action plan and cost is accurately reflected in forward workplan.

Note: When the preferred investment is selected, the focus is on identifying material risk and fatal flaws with only high levels of potential mitigation strategies. In most instances, detailed control measures are not explored at this phase and only high-level strategies should be included in the forward workplan rather than detailed actions.

Risk management tools and techniques

During the Pre-Feasibility Study, the objective is to understand the risks associated with each alternative investment option with a particular focus on identifying material risks or fatal flaws. The preferred risk assessment techniques to use are:

- *Multi-criteria analysis* – Used to compare several potential investment alternatives using scores and ratings against multiple criteria (including risk) linked to the project objectives, to identify the preferred investment alternative.
- *Option analysis* – Used to compare the relative risk of alternative options with each other and to rank options against each other.
- *SWOT analysis* – Used to understand the strengths, weaknesses, opportunities and threats of each investment alternative.
- *Scenario analysis* – Used where multiple options or possibilities are known but their probability of likelihood is unknown.

Risk lead accountabilities

- Agree to scope of risk assessment based on project objectives with project lead.
- Agree on subject matter experts and functional resources to participate in risk assessment.
- Coordinate and schedule risk workshops.
- Conduct risk workshop(s).
- Capture risk outcome in agreed risk register template.
- Agree to risk ownership, control ownership and action ownership.
- Implement process for continual review and update of risk register until end of phase.
- Obtain risk register and action sign-off from project lead team.
- Draft the risk chapter for the Pre-Feasibility Study report.
- Draft risk management plan for next phase.
- Agree on risk management strategies to go into forward workplan.

Feasibility Study

Objective

To identify effective critical controls and their performance standards, for managing material risks for the selected investment alternative.

Key deliverables

- Review, update, implement and maintain risk register and action plans consistent with the risk register.
- Continue to develop the risk register for the preferred investment option, with specific focus on further developing controls and action plans with a particular focus on the material risks.
- Produce a bow-tie diagram for each material risk identified for the preferred investment option.
    - Determine the controls, with particular focus on the critical controls, for each of the identified material risks.
    - Develop simple, practical and measurable (binary) performance standards for each of the identified critical controls. These are used to clearly demonstrate that each critical control is in place and effective.
- Review cost and schedule range analysis to support risk contingency included in execution estimate.
- Prepare the risk section for the Feasibility Study report.

- Prepare a risk management plan for the execution phase.
- Coordinate handover of risk register from the project development team to the project execution team.

Risk management tools and techniques

During the Feasibility phase the objective is to understand controls and risk treatment actions that will reduce material risk to an acceptable level. The preferred risk assessment techniques to use are:

- *Bow-tie analysis* – Used to assess risks that have causes and consequences to identify controls to prevent causes or to mitigate or minimise the consequence.
- *What if analysis* – Used to examine consequences of change to risk.
- *Hazard identification (HAZID)* – A workshop-based qualitative risk analysis technique often used for the identification of potential hazards and threats in a process, during the early design phase.
- *Hazard and operability study (HAZOP)* – Designed as an evaluation tool to identify significant hazards and ensure that there are appropriate measures to eliminate the risk or reduce risks to a tolerable level.
- *Quantitative cost risk analysis (QCRA) and schedule risk analysis (QSRA)* – Used to ensure a predictive probabilistic analysis approach to risk management. Method of risk analysis used to numerically assess the nature, sources and impacts of a risk, and assess and quantify the overall impact of uncertainties on cost or schedule, typically using Monte Carlo simulation. This is generally completed by the finance team.

Risk lead accountabilities

- Continue to maintain risk register with review and update by all project disciplines.
- Track and follow-up on the implementation of actions in the risk register.
- Obtain output from all risk techniques used during the design phase (ie FMEA, HAZOP, SiD, etc) and coordinate alignment and input into the risk register.
- Conduct assurance and control effectiveness test to ensure controls and actions are appropriately implemented to reduce the risk to an acceptable level.
- Implement process for continual review and update of risk register until end of phase.
- Obtain risk register and action sign-off from project lead team.
- Draft the risk chapter for the Feasibility Study report.
- Draft risk management plan for next phase.
- Coordinate handover of risk register from the project development team to the project execution team.

CONCLUSIONS

Appropriate risk management discipline followed through the study and project development life cycles will provide stakeholders with the required level of confidence to make informed investment decisions at key stage-gate decision points.

The level of understanding of risk will mature from selection to design phase. The key focus, from a risk perspective, for each study and project development phase is:

- *Scoping Study* – To identify material risks and project fatal flaws for a non-optimised scoping level project option.

- *Pre-Feasibility Study* – To identify risks associated with each of the Pre-Feasibility Study options being considered and use this output as a key input in the selection process for a preferred investment alternative.
- *Feasibility Study* – To identify effective critical controls and their performance standards for managing material risks for the selected investment alternative.

ACKNOWLEDGEMENTS

The processes and the majority of content provided in this chapter are derived from the risk management disciplines deployed by Auxilium One Pty Ltd in proving risk management services to development projects in the mining and resource industry. Auxilium One Pty Ltd approve the publication of this information and support and encourage the use thereof to increase and improve risk management skills across project disciplines.

FURTHER READING

International Organization for Standardization (ISO), 2018. ISO 31000:2018 Risk management – Guidelines. Available from: <https://www.iso.org/standard/65694.html>.

Shevtshenko, E, Karaulova, T and Kramarenko, S, 2008. Risk factors in project management life cycle, in *Proceedings 6th International DAAAM Baltic Conference, Industrial Engineering* (in Tallinn, Estonia).

www.ingramcontent.com/pod-product-compliance
Lightning Source LLC
LaVergne TN
LVHW071338210926
848486LV00003B/170